GROUNDWATER RESEARCH AND ISSUES

GROUNDWATER RESEARCH AND ISSUES

WILLIAM B. PORTER
AND
CHARLES E. BENNINGTON
EDITORS

Nova Science Publishers, Inc.
New York

NOTICE TO THE READER

The Publisher has taken reasonable care in the preparation of this book, but makes no expressed or implied warranty of any kind and assumes no responsibility for any errors or omissions. No liability is assumed for incidental or consequential damages in connection with or arising out of information contained in this book. The Publisher shall not be liable for any special, consequential, or exemplary damages resulting, in whole or in part, from the readers' use of, or reliance upon, this material.

Independent verification should be sought for any data, advice or recommendations contained in this book. In addition, no responsibility is assumed by the publisher for any injury and/or damage to persons or property arising from any methods, products, instructions, ideas or otherwise contained in this publication.

This publication is designed to provide accurate and authoritative information with regard to the subject matter covered herein. It is sold with the clear understanding that the Publisher is not engaged in rendering legal or any other professional services. If legal or any other expert assistance is required, the services of a competent person should be sought. FROM A DECLARATION OF PARTICIPANTS JOINTLY ADOPTED BY A COMMITTEE OF THE AMERICAN BAR ASSOCIATION AND A COMMITTEE OF PUBLISHERS.

LIBRARY OF CONGRESS CATALOGING-IN-PUBLICATION DATA
Groundwater research and issues / William B. Porter and Charles E. Bennington, editors.
    p. cm.
 Includes bibliographical references and index.
 ISBN 978-1-60456-230-9 (hardcover)
 1. Groundwater--Pollution. 2. Groundwater recharge. 3. Well water. I. Porter, William B., 1954-
 II. Bennington, Charles E.
 TD426.G767   2007
 628.1'68--dc22   2007047061

Published by Nova Science Publishers, Inc. New York

CONTENTS

PREFACE

Groundwater is water located beneath the ground surface in soil pore spaces and in the fractures of lithologic formations. A unit of rock or an unconsolidated deposit is called an aquifer when it can yield a usable quantity of water. The depth at which soil pore spaces or fractures and voids in rock become fully saturated with water is called the water table. Groundwater is recharged from, and eventually flows to, the surface naturally; natural discharge often occurs at springs and seeps, streams and can form oases or wetlands. Groundwater is also often withdrawn for agricultural, municipal and industrial use by constructing and operating extraction wells. The study of the distribution and movement of groundwater is hydrogeology, also called groundwater hydrology.

Typically groundwater is thought of as liquid water flowing through shallow aquifers, but technically it can also include soil moisture, permafrost (frozen soil), immobile water in very low permeability bedrock, and deep geothermal or oil formation water. Groundwater is hypothesized to provide lubrication which can possibly aid faults to move. This new book presents important research in the field.

Chapter 1 - In this chapter, experimental tests of different techniques for contaminated soils and groundwater remediation are presented and discussed. The contaminant of concern is the MtBE, a common antiknock additive in the production of unleaded fuel. MtBE's great water solubility, low partitioning into soils and low tendency to be degraded by microbial activity, are responsible for a considerable mobility that allows rapid dispersion and strong persistency in groundwater.

To remove or/and destroy this contaminant, three *in situ* technologies are evaluated and compared: Air Sparging (AS), soil flushing (ISSF), and chemical oxidation (ISCO). AS, which mainly involves the stripping of the contaminant by injected air in the subsurface, is currently one of the most commonly used, due to the simplicity of the plant installation, operation and maintenance and, consequently, to its low associated costs. Other systems, like ISSF and ISCO can be also successfully implemented in MtBE contaminated soils and groundwater treating. Both ISSF and ISCO technologies deal with the chemical oxidation of the pollutant, with the difference that for ISSF the oxidant does not come into contact with the soil matrix. In the experimentation presented in this chapter, Fenton oxidation was performed in either technologies used.

For each technology after the description of the basic principles and environmental applications, the results of lab scale experiments conducted on samples of saturated soil artificially contaminated, are discussed.

AS tests were performed both in a cylindrical column (small scale test) and in a large tank of 1 m^3 (intermediate scale test). These two testing set-ups were used to study the optimal air flow rate for MtBE transfer to the gas phase and the influence of the geometry around the injection well.

ISSF tests were performed by flushing the soil with an aqueous solution of a surfactant agent (ethanol). The pH, the ethanol concentration and the flow rate of this extracting solution were optimized to achieve the complete removal of MtBE from the matrix. The recovered solution was then oxidized with Fenton's reagent to destroy the pollutant. The effect of ethanol concentration towards MtBE oxidation was also evaluated, considering the non selectivity of the oxidant.

ISCO tests were performed by injecting Fenton's reagent directly into the soil, both at the natural pH of soil and under acidic conditions.

Chapter 2 - Land subsidence due to excessive extraction of groundwater is a major environmental problem that has affected for several decades both the urban centers and the surrounding peri-urban farmlands scattered within the Nobi Plain, Central Japan. Even though strict regulations successfully reduced the subsidence rates, the threat is still far from being eliminated. The plain is one of the most important economic centers of Japan, sustaining an intensive agricultural and industrial production. Furthermore, its population density is among the highest in the country, averaging more than 1,000 persons per km^2. The recent economical growth of the region is expected to be accompanied by new developments and a higher demand for water supply; therefore, authorities are compelled to update management strategies to satisfy the increasing demand without causing undesirable effects on the environment. Knowledge of the groundwater availability becomes an essential requisite for the establishment of these policies. A numerical simulation was applied to understand the groundwater flow in the region. The resulting model was then used to determine the safe yield of the aquifers under different management options. The analysis concentrated on the Aburashima site, as this area faces the possibility of development in the forthcoming years. Calculations indicated that withdrawal from the shallow aquifer is limited. In addition, a decline in piezometric levels reduced the groundwater availability in summer. Spreading a number of wells around the site constitutes an alternative to increase the pumping volume. Higher quantities and lower extraction costs makes the lower aquifer a more reliable source of groundwater. The present study constitutes one of the first steps towards the implementation of an integrated management plan of the water resources in the Nobi Plain, and constitutes a valid reference to update the current legislation regarding groundwater use in the region.

Chapter 3 - Most large urban centers lie in coastal regions, which are home to about 25% of the world's population. The current coastal urban population of 200 million is projected to almost double in the next 20 to 30 years. This expanding human presence has dramatically changed the coastal natural environment. To meet the growing demand for more housing and other land uses, land has been reclaimed from the sea in coastal areas in many countries, including China, Britain, Korea, Japan, Malaysia, Saudi Arabia, Italy, the Netherlands, and the United States. Coastal areas are often the ultimate discharge zones of regional ground water flow systems. The direct impact of land reclamation on coastal engineering, environment and marine ecology is well recognized and widely studied. However, it has not been well recognized that reclamation may change the regional groundwater regime, including groundwater level, interface between seawater and fresh groundwater, and submarine groundwater discharge to the coast.

This paper will first review the state of the art of the recent studies on the impact of coastal land reclamation on ground water level and the seawater interface. Steady-state analytical solutions based on Dupuit and Ghyben-Herzberg assumptions have been derived to describe the modification of water level and movement of the interface between fresh groundwater and saltwater in coastal hillside or island situations. These solutions show that land reclamation increases water level in the original aquifer and pushes the saltwater interface to move towards the sea. In the island situation, the water divide moves towards the reclaimed side, and ground water discharge to the sea on both sides of the island increases. After reclamation, the water resource is increased because both recharge and the size of aquifer are increased.

This paper will then derive new analytical solutions to estimate groundwater travel time before and after reclamation. Hypothetical examples are used to examine the changes of groundwater travel time in response to land reclamation. After reclamation, groundwater flow in the original aquifer tends to be slower and the travel time of the groundwater from any position in the original aquifer to the sea becomes longer for the situation of coastal hillside. For the situation of an island, the water will flow faster on the unreclaimed side, but more slowly on the reclaimed side. The impact of reclamation on groundwater travel time on the reclaimed side is much more significant than that on the unreclaimed side. The degree of the modifications of the groundwater travel time mainly depends on the scale of land reclamation and the hydraulic conductivity of the fill materials.

Chapter 4 - Groundwater numerical models are powerful and efficient tools for groundwater management, protection, and remediation. However, groundwater modelling, which requires a huge amount of data, is not an easy endeavor. To build a predictive model, and to get reliable results, input data should be accurate and representative of the real situation in the field. Because of the randomness inherent in nature and the heterogeneity of aquifers, it is very difficult to accurately determine the hydrological properties of the aquifers.

Classical groundwater models usually handle input parameters in a deterministic way, without considering any variability, uncertainty, or randomness in these parameters. Thus, the results of deterministic modelling are questionable.

To account for uncertainty in physical, chemical, and geological data, stochastic modelling is usually used. Many approaches have been developed and used, including sampling approaches, reliability methods, and the Monte Carlo simulation. In this chapter, different approaches of stochastic and probabilistic modelling are introduced and discussed.

Chapter 5 - While TCE and perchlorate are both classified by DOD as emerging contaminants, there are important distinctions in how they are regulated and in what is known about their health and environmental effects. Since 1989, EPA has regulated TCE in drinking water. However, health concerns over TCE have been further amplified in recent years after scientific studies have suggested additional risks posed by human exposure to TCE. Unlike TCE, no drinking water standard exists for perchlorate—a fact that has caused much discussion in Congress and elsewhere. Recent Food and Drug Administration data documenting the extent of perchlorate contamination in the nation's food supply has further fueled this debate.

While DOD has clear responsibilities to address TCE because it is subject to EPA's regulatory standard, DOD's responsibilities are less definite for perchlorate due to the lack of such a standard. Nonetheless, perchlorate's designation by DOD as an emerging contaminant has led to some significant control actions. These actions have included responding to requests by EPA and state environmental authorities, which have used a patchwork of statutes, regulations, and general oversight authorities to address perchlorate contamination. Pursuant to its Clean Water Act authorities, for example, Texas required the Navy to reduce perchlorate levels in wastewater discharges at the McGregor Naval Weapons Industrial Reserve Plant to 4 parts per billion (ppb), the lowest level at which perchlorate could be detected at the time. In addition, in the absence of a federal perchlorate standard, at least nine states have established nonregulatory action levels or advisories for perchlorate ranging from 1 ppb to 51 ppb. Nevada, for example, required the Kerr-McGee Chemical site in Henderson to treat groundwater and reduce perchlorate releases to 18 ppb, which is Nevada's action level for perchlorate.

While nonenforceable guidance had existed previously, it was not until EPA adopted its 1989 TCE standard that many DOD facilities began to take concrete action to control the contaminant. According to EPA, for example, 46 sites at Camp Lejeune have since been identified for TCE cleanup. The Navy and EPA have selected remedies for 30 of those sites, and the remaining 16 are under active investigation. Regarding perchlorate, in the absence of a federal standard DOD has implemented its own policies on sampling and cleanup, most recently with its 2006 *Policy on DOD Required Actions Related to Perchlorate*. The policy applies broadly to DOD's active and closed installations and formerly used defense sites within the United States and its territories. It requires testing for perchlorate and certain cleanup actions and directs the department to comply with applicable federal or state promulgated standards, whichever is more stringent. The policy notes, that DOD has established 24 ppb as the current level of concern for managing perchlorate until the promulgation of a formal standard by the states and/or EPA.

Chapter 6 - DOD has implemented or field-tested all of the 15 types of generally accepted technologies currently available to remediate contaminated groundwater, including several alternatives to pump-and-treat technologies. Some of these technologies, such as bioremediation, introduce nutrients or other materials into the subsurface to stimulate microorganisms in the soil; these microorganisms consume the contaminant or produce byproducts that help break down contaminants into nontoxic or less-hazardous materials. DOD selects the most suitable technology for a given site on the basis of several factors, such as the type of contaminant and location in the subsurface, and the relative cost-effectiveness of a technology for a given site. DOD has identified a number of contaminants of concern at its facilities, each of which varies in its susceptibility to treatment. The table below shows the technologies DOD used to remediate contaminated groundwater.

GAO did not identify any alternative groundwater remediation technologies being used or developed outside DOD that the department has not considered or used. Most of the new approaches developed by commercial vendors and available to DOD generally use novel materials applied to contaminated sites with existing technologies. DOD actively researches and tests new approaches to groundwater remediation largely by developing and promoting the acceptance of innovative remediation technologies. For example, DOD's Strategic Environmental Research and Development Program supports public and private research on contaminants of concern to DOD and innovative methods for their treatment.

Chapter 7 - In a multi-year study, the Clean Air Task Force (CATF) examined 15 coal mines where coal ash was placed under the Pennsylvania Department of Environmental Protection (PADEP) Coal Ash Beneficial Use Program, which encourages the placement of coal combustion waste (CCW) in active and abandoned mines. The study concludes that the state's beneficial use program, whose primary goal is to improve the environmental condition of mines by adding of massive quantities of CCW, is failing:

- At 10 of the 15 minefills examined in the study, monitoring data indicate the coal ash contaminated groundwater or streams.
- At three minefills contamination of streams and/or groundwater was occurring, but the cause of the pollution could not definitively be traced to the ash because of the lack of monitoring data.
- At one of the minefills, water quality improved for acid mine parameters, but water quality decreased for contaminants found in ash.
- At only 1 of the 15 minefills was water quality improvement found. However, since water monitoring was terminated shortly after placement, it is unknown whether the improvement was temporary.

Consequently, CATF found that placing large amounts of CCW in mines is a dangerous practice that appears to be causing toxic levels of contamination. The report recommends that permits allowing this industrial waste to be placed in mines require safeguards to minimize adverse Environmental impacts and threats to human health.

CCW is a toxic industrial waste produced by coal-burning power plants. Pennsylvania is the third largest US producer of this waste, generating over 9 million tons per year. CCW contains hazardous chemicals including aluminum, chloride, iron, manganese, sulfate and toxic trace elements such as arsenic, selenium, lead, mercury, cadmium, nickel, copper, chromium, boron, molybdenum and zinc.

For over 20 years, PADEP has been promoting placement of large volumes of CCW in active and abandoned coal mines to address acid mine drainage, increase soil fertility, and fill mine pits and voids. PADEP has permitted approximately 120 CCW minefills throughout the state.

Chapter 8 - Remedial technologies utilized at hazardous waste sites for the treatment of metal and metalloid contaminants often take advantage of reduction-oxidation (redox) processes to reach ground water clean up goals. This is because redox reactions, in many cases, govern the biogeochemical behavior of inorganic contaminants by affecting their solubility, reactivity, and bioavailability. Site characterization efforts, remedial investigations, and long-term post-remedial monitoring often involve sampling and analysis of solids. Solid-phase studies are needed to evaluate contaminant partitioning to various mineral fractions, to develop site conceptual models of contaminant transport and fate, and to assess whether or not remedial mechanisms are occurring as expected. Measurements to determine mineralogical compositions, contaminant-mineral associations, and metal/metalloid uptake capacities of subsurface solids or reactive media used for *in situ* treatment of the subsurface all depend upon proper sample collection and preservation practices. This Issue Paper discusses mineralogical preservation methods for solid samples that can be applied during site characterization studies and assessments of remedial performance. A preservation protocol is presented that is applicable to solids collected from anoxic subsurface environments, such as soils, aquifers, and sediments.

The preservation method evaluated and recommended here for solids collected from anoxic environments involves sample freezing (-18 °C), transportation of frozen samples on dry ice, and laboratory processing of solids in an anaerobic glove box. This method was found to preserve the redox integrity of reduced iron- and sulfur-bearing compounds, which are typically predominant redox-sensitive inorganic constituents in environmental materials andareimportantincontrollingcontaminantbehaviorathazardous waste sites. A selection of solid-phase measurements was carried out on preserved anoxic sediments collected from a contaminated lake and compared to identical measurements on sample splits in which no preservation protocol was adopted, i.e., the unpreserved samples were allowed to oxidize in ambient air. An analysis of results illustrates the importance of proper sample preservation for obtaining meaningful solid-phase characterization. This paper provides remedial project managers and other state or private remediation managers and their technical support personnel with information necessary for preparing sampling plans to support site characterization, remedy selection, and post-remedial monitoring efforts.

In: Groundwater Research and Issues   ISBN: 978-1-60456-230-9
Editors: W. B. Porter, C. E. Bennington, pp. 1-28   © 2008 Nova Science Publishers, Inc.

Chapter 1

IN SITU TECHNOLOGIES FOR HAZARDOUS MATERIALS-CONTAMINATED SOILS AND GROUNDWATER: THE CASE OF MtBE

L. di Palma[1], C. Alimonti[1], D. Lausdei[2] and R. Mecozzi[1]*
[1]Università di Roma "La Sapienza", Dip. Ingegneria Chimica, Roma, Italy
[2]Golder Associates s.r.l., Roma, Italy

ABSTRACT

In this chapter, experimental tests of different techniques for contaminated soils and groundwater remediation are presented and discussed. The contaminant of concern is the MtBE, a common antiknock additive in the production of unleaded fuel. MtBE's great water solubility, low partitioning into soils and low tendency to be degraded by microbial activity, are responsible for a considerable mobility that allows rapid dispersion and strong persistency in groundwater.

To remove or/and destroy this contaminant, three *in situ* technologies are evaluated and compared: Air Sparging (AS), soil flushing (ISSF), and chemical oxidation (ISCO). AS, which mainly involves the stripping of the contaminant by injected air in the subsurface, is currently one of the most commonly used, due to the simplicity of the plant installation, operation and maintenance and, consequently, to its low associated costs. Other systems, like ISSF and ISCO can be also successfully implemented in MtBE contaminated soils and groundwater treating. Both ISSF and ISCO technologies deal with the chemical oxidation of the pollutant, with the difference that for ISSF the oxidant does not come into contact with the soil matrix. In the experimentation presented in this chapter, Fenton oxidation was performed in either technologies used.

For each technology after the description of the basic principles and environmental applications, the results of lab scale experiments conducted on samples of saturated soil artificially contaminated, are discussed.

---

* corresponding author: luca.dipalma@uniroma1.it, tel: +39-06-44585571; fax: +39-06-44585622

AS tests were performed both in a cylindrical column (small scale test) and in a large tank of 1 m^3 (intermediate scale test). These two testing set-ups were used to study the optimal air flow rate for MtBE transfer to the gas phase and the influence of the geometry around the injection well.

ISSF tests were performed by flushing the soil with an aqueous solution of a surfactant agent (ethanol). The pH, the ethanol concentration and the flow rate of this extracting solution were optimized to achieve the complete removal of MtBE from the matrix. The recovered solution was then oxidized with Fenton's reagent to destroy the pollutant. The effect of ethanol concentration towards MtBE oxidation was also evaluated, considering the non selectivity of the oxidant.

ISCO tests were performed by injecting Fenton's reagent directly into the soil, both at the natural pH of soil and under acidic conditions.

INTRODUCTION

Methyl *tertiary* butyl ether (MtBE) is a volatile, flammable, uncolored transparent liquid with a strong hydrocarbon's smell.

This product, which has been synthesized since the early 1970s (as an antiknock additive for unleaded fuel, instead of tetraethyl lead), had a dramatic increase of production during the 1990s, when in some areas of the U.S. the use of oxygenates was required in common gasoline formulations to improve air quality, by improving the combustion of the volatile compounds, reducing the heat of vaporisation and lowering the boiling point of gasoline.

The principal chemical and physical properties of MtBE are:

- Molecular weight (g/mol): 88.15
- Relative vapor density (air=1): 3.0
- Relative liquid density (water=1): 0.744
- Vapor tension (mmHg): 245-276 at 25°C
- Henry's constant : 0.02-0.05
- Water solubility (mg/l): 43000-54000 at 25°C
- Boiling Point (°C): 55.2
- Melting Point (°C): -109
- log Koc: 1.035-1.091
- log Kow: 1.24

The polarity of the ether group in the molecule enables it to have any sort of hydrogen bond with water.

Despite the fact that MtBE was mainly introduced for environmental purposes, in the last decade groundwater contamination by this product due to fuel spills or leakage from underground storage facilities became a highly sensitive issue. When it reaches the soil, this pollutant partially volatilizes and is partially absorbed. Unfortunately, due to its great water solubility and low biodegradability, it has a great mobility that allows a rapid dispersion and contamination of the groundwater, especially after rainfall. It is therefore very persistent in the environment, and the plume generated by MtBE contamination is generally of a greater extension than that of other pollutants that may be present in the fuel.

As a consequence, the need exists to remove MtBE from the contaminated environments.

The technologies commonly used for the remediation of contaminated soil and groundwater are generally classified as:

- in-site (or *in situ*), when the contaminated environmental matrix (soil or groundwater) is treated directly in place, without pulling it out of the subsoil;
- on-site, when the contaminated matrix is treated on the surface, directly above the place where it has been extracted;
- off-site (or *ex situ*), when the contaminated matrix is treated away from the site where it has been extracted.

Generally, *in situ* technologies are cheaper and safer than the *ex situ* ones, because excavation and management of huge quantities of material are not required. As a consequence, the related risk of pollutant spread in the environment is also reduced.

In this chapter, applications for the remediation of soil and groundwater contaminated by MtBE by a physical treatment, the air sparging (AS) technology, and two chemical treatments, soil flushing (ISSF) and chemical oxidation (ISCO), are described and experimentally evaluated. After a general description of the principles of each technology, a summary of experimental research at lab-scale and intermediate pilot scale are presented and discussed.

AIR SPARGING

Air sparging (AS) is an *in situ* technology designed to remove volatile organic compounds (VOCs) from the saturated zone through the injection of a gas under pressure (generally air) from one or more wells (specifically designed for injection) below the contaminated plume. AS has been used to address a broad range of volatile and semi-volatile groundwater and soil contaminants, including gasoline and other fuels, associated components such as BTEX or MtBE, and chlorinated solvents (Miller R.R., 1996).

The target of the air injection is to volatilize groundwater contaminants and to promote biodegradation in saturated and unsaturated zone, by increasing subsurface oxygen concentration. Volatilized vapours migrate into vadose zone where they are extracted via vacuum, generally by a soil vapour extraction (SVE) system (Figure 1).

The injected air travels upwards through buoyancy as discrete bubbles or in channels. This produces the mass transfer of the dissolved VOCs from the liquid phase into the vapour phase through the air-water interfacial surface created in the soil pore space (stripping process). AS applicability is restricted to unconfined aquifers where the injected air, extracting organic vapours, can migrate to the unsaturated zone through the water surface.

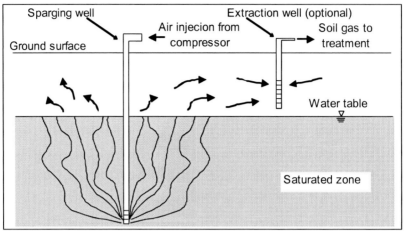

Figure 1. Scheme of an Air Sparging system coupled with a SVE well for the organic vapors recovery

The removal effectiveness of AS technology highly depends on the geological characteristics of the soil as well as on the physico-chemical properties of the polluting compounds (Burns and Zhang, 2001). In effect, AS effectiveness relies on the fact that the airflow could reach all the volume of aquifer to treat. Thus, the most suitable aquifers to be treated by AS technology consist of permeable rock or soils ranging from medium to highly permeable (aquifers with values of the *hydraulic conductivity* within the range: $10^{-4} < K < 10^{-2}$ m/s). With respect to the physical-chemical properties of the pollutants, it is mandatory that they allow a quick mass transfer into the vapour phase within the air volumes (bubbles or channels); hence, the most suitable chemicals are the ones classified as volatile and semi-volatile (Lausdei, 2003).

IN SITU SOIL FLUSHING

In situ soil flushing performs the removal of organic or inorganic pollutants from soil, through the infiltration of an extracting fluid in the subsurface and its successive extraction together with the groundwater.

The flushing fluid is typically an aqueous solution and may contain many additives, such as chelating agents or surfactants that, due to their particular molecular structure, can help the extraction of the contaminant from the aqueous phase. Surfactants in particular, have a hydrophilic head and an oleophilic tail, which allow them to alter the properties of organic-water interfaces (Jafvert 1996; Roote 1998). The retention of contaminants by soil is affected by many factors. Among them the most important are: the soil nature, the pH, the soil cation-exchange capacity and the eventual simultaneous presence of different organic pollutants. Hydraulic conductivity (K) is the principal physical parameter that can determine the feasibility of a soil flushing process (Freeman and Harris, 1995). In effect, as soils with low permeability ($K < 1.0 \cdot 10^{-5}$ cm/s) naturally inhibit the permeation of fluids, soil flushing proves to be effective only for those classified as permeable ($K > 1.0 \cdot 10^{-3}$ cm/s) or, to a lesser extent, as slightly permeable ($1.0 \cdot 10^{-5} < K < 1.0 \cdot 10^{-3}$ cm/s).

In situ soil flushing technologies must be combined with the successive treatment of the extracted solution (Pump & Treat). The mixture of the injected fluid and contaminant (the extracted solution), along with ground water, is collected through extraction wells and sent to the successive chemical treatment (Freeman and Harris 1995; Di Palma et al. 2003). This is done to destroy the pollutants or to remove them from the liquid, allowing their recycle in the process and the discharging of all treated water, which can diminish the overall treatment costs.

The flushing technology avoids the direct contact of the soil with the oxidant. However it produces a large amount of effluent to be treated and needs a lot of time to achieve the results. In addition, though the contact time of the flushing solution with the soil is generally low (on the contrary of the washing processes), the extracted solution will probably contain a substantial amount of soil constituents other than the target compound. These are mostly from natural organic matter, and are strongly susceptive to the successive oxidation. Furthermore, since the surfactant itself can undergo oxidation, the process needs a previous separation stage to recover it. This permits a considerable reduction in chemical reagents consumption. To reduce the extracted solution's volume, membrane filtration, liquid extraction or evaporation processes are necessary, thus increasing the overall cost of the treatment.

The *in situ* flushing technology (extraction, followed by oxidation), has been already supported by decades of experimental laboratory, pilot and field tests. Therefore, it is commonly adopted in the remediation of contaminated soils. In the remediation of heavy metals polluted soil, the effectiveness of EDTA (ethylenediaminetetraacetic acid, E-H$_4$) as a chelating agent in flushing treatments has been widely demonstrated (Beckett, 1989; Tejowulan, 1998; Martin and Allen, 1996). Using EDTA presents some advantages in comparison with other chelating agents: it has a low degree of biodegradability in soil and groundwater, coupled with a high capacity of complexing heavy metals (Sun et al., 2001). Conversely, the high consumption and cost of EDTA, strongly affects the economic feasibility of a soil flushing remediation (Chang, 1995; Di Palma et al., 2003).

In the remediation of organic pollutants contaminated soils, the addition of flushing additives, as surfactants, to the extracting solution has also proven to be highly effective (Beckett, 1989; Tejowulan, 1998; Liu et al., 1995; Chang et al., 2000).

The implementation of the flushing technology involves the design and installation of a system of injection and extraction wells.

IN SITU CHEMICAL OXIDATION

The chemical oxidation of contaminated soils or groundwater can be performed through *ex situ* or *in situ* technologies.

The *ex situ* chemical oxidation performs the mixing of the oxidizing agent with soil or groundwater in a vessel. Excavation and transportation of the soil are therefore necessary: those operations not only largely affect the overall cost of the treatment, but also may imply additional risks of environmental contamination (Ghassemi 1988). The oxidizing agent can be an aqueous solution, or less commonly, a gas (ozone). The most frequently used oxidizing agents for the treatment of organic contaminants are: ozone, hydrogen peroxide and potassium permanganate.

In the past few years (US EPA 1998; Siegrist 1998; Yin and Allen, 1999; ITRC 2000; Chen et al. 2001; Quan et al. 2003), chemical oxidation processes have been successfully introduced in the remediation of contaminated soils. In recent times, *in situ* chemical oxidation technology (direct oxidation) has become to be implemented due to its potential advantages, both as a unique remediation technology, and as a pretreatment step to enhance biodegradation. Its implementation is particularly useful in case of toxic or recalcitrant compounds because it's aggressive and non selective.

In situ chemical oxidation involves the direct injection of chemical oxidants into the vadose zone or groundwater (Fig. 2).

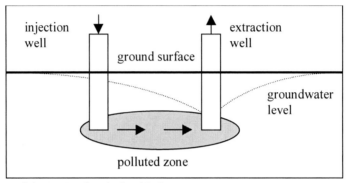

Figure 2. Scheme of the *in situ* chemical oxidation process.

The objective of the treatment can be the total degradation of the contaminant or simply its transformation in an oxidized product which can undergo to a successive biological degradation (Ghassemi, 1988). The performance of the treatment is greatly affected by the distribution of the contamination, and by soil characteristics such as pH, permeability, and the presence of other matter susceptive to oxidation (ITRC 2001; Tyre et al. 1991; Tarr and Lindsay 1998).

The main advantage of this emerging technology towards the *ex situ* one, is that soil excavation and mobilization are avoided, and consequently the risks of release of contaminants into the wider environment are reduced (Di Palma, 2005).

PHYSICAL TREATMENT

Experimental Tests of MtBE Remediation by Air Sparging

Laboratory Scale Experiments

In this section the results of experimentation on the MtBE removal from a contaminated aquifer, performed by Air Sparging (AS), are presented and discussed. The objective was to evaluate the effectiveness of AS in MtBE removal from a saturated porous media. Beside this, tests were performed also to evaluate and compare two types of air diffusers, each with a different orifice size.

The laboratory apparatus used to perform experimental tests consisted of the following components, as shown in Figure 3:

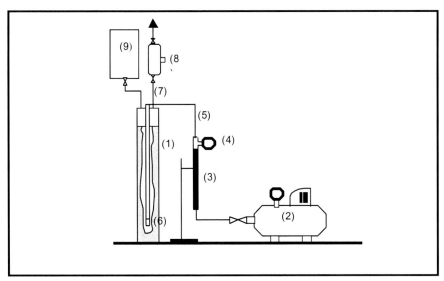

Figure 3. Laboratory experimental apparatus and related flow diagram
(1) plexiglas column (11 x 4.5 x 100 cm), with a top wooden cap, packed with a saturated porous media;
(2) air compressor, to provide pressurized air to the system;
(3) flowmeter, to measure the injected airflow;
(4) manometer, to measure the injected air pressure;
(5) "Rilsan" line, to join the sparge point to the air compressor; and
(6) sparge point, to inject air within the saturated media.
(7) vapour recovery line, to drive vapours out of the column;
(8) vapour sampler (250 ml), to withdraw vapour samples coming out of the column; and
(9) saturation tank, to prepare and store the contaminant solution.

The column was packed with highly uniform gravel and glass beads (idealized porous media), in order to easily replicate identical conditions for each test.

The first sampling point, conventionally called A, was placed at 50 cm above the bottom of the column, the second point, conventionally called B, was placed at the bottom of the column below the zone reached by injected air.

The sparge point was placed within the column, just at 10 cm above the bottom, to allow the injected airflow to have enough space for its developing upwards. In table 1 the main characteristics of the two diffusers tested are reported.

Table 1. Diffusers' geometrical properties

Diffuser no.	Inside diameter (mm)	Outside diameter (mm)	Orifice size (μm)
1	4	8	<1.6
2	4	8	41÷100

The column was saturated with a solution of MtBE in water, at the concentration of 10 mg/l. The solution was prepared inside the saturation tank and then poured within the column through nylon piping. Using a peristaltic pump, three continuous cycles of saturation and drainage were completed to uniformly spread the solution within the saturated media.

The removal tests began with air injection start-up, done right after the initial sampling of solution, and were pursued with several other scheduled samplings of solution.

Results of the two removal tests are shown in Figure 4. The reported concentrations of MtBE are normalized with their initial concentrations, in order to compare the results of both diffusers in terms of removal percentage.

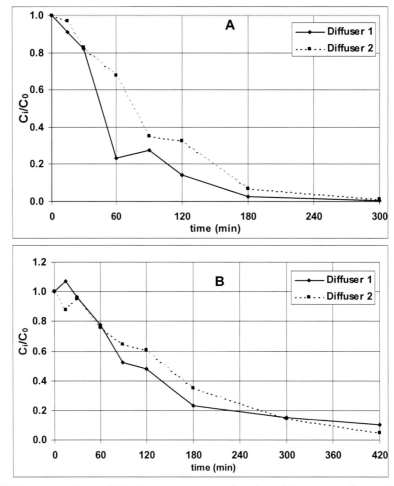

Figure 4. MtBE concentrations within the pore water as a function of time at sampling points A and B.

Comparing the results of the tests obtained using diffusers 1 and 2, it can be noticed that:

- with both diffusers, the residual concentration of MtBE decreases more rapidly at the point A because this is directly impacted by the injected airflow and therefore the stripping process is stronger than at point B; and
- at both sampling points, the concentration of MtBE decreases more rapidly when air is injected using diffuser 1, the one which produced the highest percentage of smaller size bubbles as evidenced during a previous image analysis tests.

In both cases, the results showed that all the MtBE initially dissolved in the pore water solution, was stripped by the air flow injected within the saturated media and passed in the vapour phase. Therefore the treatment proved to be effective.

Intermediate Scale Experiments

MtBE removal from a contaminated aquifer by AS was also performed in a intermediate scale test. This was carried out in the experimental apparatus described in figure 5, consisting in a steel tank (dimensions 1 x 1 x 1,2 m) containing 1 m^3 of soil (Alimonti et al., 2006). Figure 6 shows the sampling points distribution within the saturated porous media.

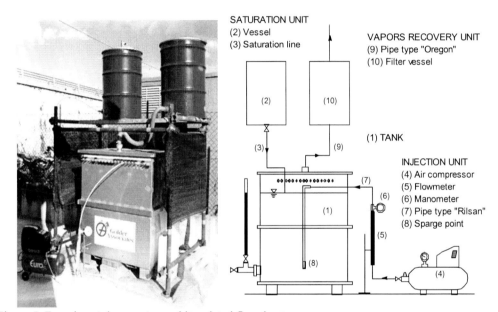

Figure 5. Experimental apparatus and its related flowsheet.

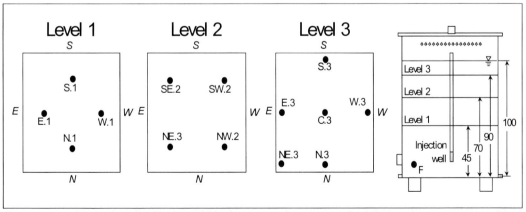

Figure 6. Sampling points distribution within the saturated porous media (all elevations are expressed in cm).

The trends of MtBE concentration observed in the experiments, are illustrated in Figure 7.

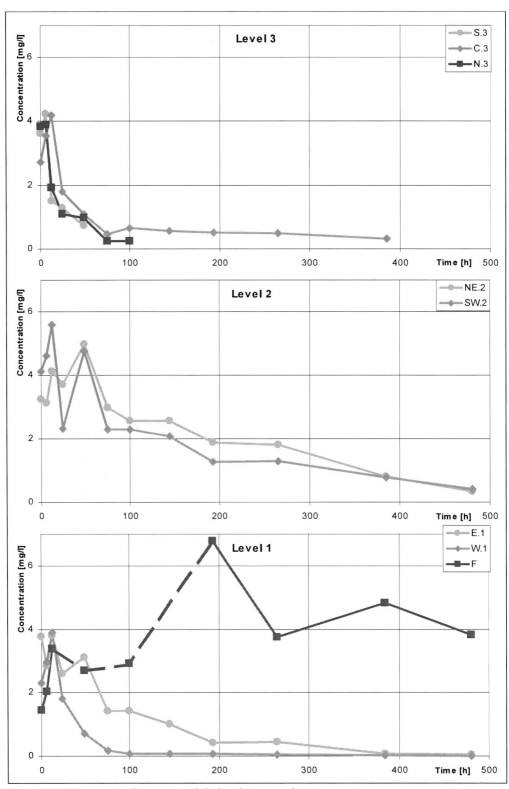

Figure 7. MtBE concentration measured during the removal test.

A general effectiveness of the AS technology in MtBE removal from the contaminated saturated media was observed whereas the media had been directly reached by the injected airflow, MtBE's concentration trend, after an initial fluctuation, shows a strong tendency to attenuate. The fluctuations of the concentration occur at the early stage of air injection, generally within the first two days of operation. After this lapse of time, the concentrations gradually decrease and there is an asymptotic tendency to zero.

Among the three main contaminant removal mechanisms (stripping, direct volatilization and aerobic biodegradation) generally involved in air sparging treatments,(Suthersan S. S., 1999), the experiment highlighted the action of stripping as the mechanism of MtBE removal.

Assuming *stripping* as the dominant mass removal mechanism, Henry's law provides a basic explanation of the trend and the distribution of MtBE concentration within the saturated media.

Since the partitioning of the dissolved contaminant mass occurs at the discrete air-water interface created by the air injection (bubbles and/or air channels), lower MtBE concentrations were expected in the immediate vicinity of the air paths, within the saturated media. Furthermore, to replenish the mass loss around the air paths, mass transfer by diffusion occurred from the water located far from air-water interfaces. Therefore, it is likely that the density of air paths played a significant role in mass transfer effectiveness, as well as the residence time of the injected air traveling through these discrete paths. The dissolved MtBE concentrations, measured at different times and sampling points, consequently rely upon local availability of interfacial surface area and upon MtBE's concentration in the vapor phase within the air paths.

CHEMICAL TREATMENTS

Principles of Chemical Oxidation with Fenton's Reagent

In case of organic contamination, hydrogen peroxide is one of the most used chemical oxidants because it is environmentally safe and easy to handle. Its high oxidizing power may be further increased through the combination with UV radiation (Maillard et al. 1992; Venkatadri and Peters 1993) or transition metal salts (Lu et al. 1997; Weiss 1935; Zepp et al. 1992). The significant increase in hydrogen peroxide's oxidizing capacity is due to the formation of the hydroxyl radical which is highly unstable but has an oxidation power ($E° = 2.8$ V), lower only to that observed for fluorine, as shown in Table 2.

The most common environmental application of hydrogen peroxide is the Fenton's reagent, an aqueous mixture of H_2O_2 and Fe^{2+} (Fenton, 1894). Under acidic condition, the reaction between iron and hydrogen peroxide generates hydroxyl radicals, which are strong enough to oxidize most organic and inorganic compounds in a non-selectively way. The following equations (Mohanty and Wei 1993) describe the reaction mechanism of the Fenton's system:

$$Fe^{2+} + H_2O_2 \rightarrow Fe^{3+} + HO^{\bullet} + OH^{-} \qquad (1)$$

The initiation reaction (1) is the reaction that leads to the hydroxyl radical generation, and it is the process limiting step (Mohanty and Wei 1993).

Table 2. Oxidation powers of oxidizing agents

Oxidizing agent	Oxidation reaction	E^0 (V)
Oxygen	$O_2 + 4\,H^+ + 4\,e^- \Leftrightarrow 2\,H_2O$	+ 1.229
Dichromate	$Cr_2O_7^{2-} + 14\,H^+ + 6\,e^- \Leftrightarrow 2\,Cr^{3+} + 7\,H_2O$	+ 1.33
Hypochlorite	$HOCl + H^+ + 2\,e^- \Leftrightarrow Cl^- + H_2O$	+ 1.495
Permanganate	$MnO_4^- + 8\,H^+ + 5\,e^- \Leftrightarrow Mn^{2+} + 4\,H_2O$	+ 1.51
Hydrogen peroxide	$H_2O_2 + 2\,H^+ + 2\,e^- \Leftrightarrow 2\,H_2O$	+ 1.776
Ozone	$O_3 + 2\,H^+ + 2\,e^- \Leftrightarrow O_2 + H_2O$	+ 2.07
Hydroxyl radical	$HO^\bullet + H^+ + e^- \Leftrightarrow H_2O$	+ 2.8
Fluorine	$F_2 + 2\,e^- \Leftrightarrow 2\,F^-$	+ 2.866

The substrate can react with the oxidants through reactions (2-4).

$$RH + HO^\bullet \rightarrow H_2O + R^\bullet \tag{2}$$
$$R^\bullet + HO^\bullet \rightarrow ROH \tag{3}$$
$$R^\bullet + H_2O_2 \rightarrow ROH + HO^\bullet \tag{4}$$

A lot of secondary reactions occur simultaneously to reaction (1):

$$Fe^{3+} + HO_2^\bullet \rightarrow Fe^{2+} + H^+ + O_2 \tag{5}$$
$$Fe^{3+} + H_2O_2 \rightarrow Fe^{2+} + HO_2^\bullet + H^+ \tag{6}$$
$$Fe^{2+} + HO^\bullet \rightarrow Fe^{3+} + OH^- \tag{7}$$
$$H_2O_2 + HO^\bullet \rightarrow H_2O + HO_2^\bullet \tag{8}$$
$$Fe^{2+} + HO_2^\bullet \rightarrow Fe^{3+} + HO_2^- \tag{9}$$

These reactions (5-9) deal with the cycling of Fe^{3+} to Fe^{2+}, with the production of hydroperoxyl radicals, whose oxidizing power is a lot less than the one of the parent compounds ($E° = 0,3$ V), and with the quenching of $HO^\bullet$ by Fe^{2+} and H_2O_2.

The presence of the substrate can also help the regeneration of Fe^{2+} through reaction (10) or its depletion through reaction (11).

$$R^\bullet + Fe^{3+} \rightarrow R^+ + Fe^{2+} \tag{10}$$
$$R^\bullet + Fe^{2+} \rightarrow R^- + Fe^{3+} \tag{11}$$

Finally, the radical chain reactions may also be ended by the reactions (12-13)

$$2\,HO^\bullet \rightarrow H_2O_2 \tag{12}$$
$$2\,R^\bullet \rightarrow dimer \tag{13}$$

The combined system of hydrogen peroxide coupled with a bivalent iron salt, results in the great number of reactions above described. As shown, numerous competing reactions manage to scavenge the hydroxyl radicals, thus decreasing the system's oxidizing power: the production of hydroxyl radical (HO$^\bullet$) is always accomplished by the production of hydroperoxyl radicals (HO$_2^\bullet$), Fe^{3+} and O$_2$.

A particular attention must therefore be given in finding the correct dosage for the reactants. In particular an excess in hydrogen peroxide leads to the formation of hydroperoxyl radicals HO$_2^\bullet$ (reaction 8), which are competitive with HO$^\bullet$ in the oxidation process, and show a certain reactivity towards organic compounds, even if they exhibit a very low oxidation power. The HO$_2^\bullet$ radicals may also oxidize Fe^{2+} ions (reaction 9), and, by subtracting them from the reaction site, may cause a deficit in HO$^\bullet$ production.

An additional, but non negligible consideration, can be made about the need of establishing acidic conditions to optimize Fenton's reaction effectiveness.

The overall Fenton reaction process can be simplified in the following reaction (Walling, 1975):

$$Fe^{2+} + H_2O_2 + 2\,H^+ \rightarrow Fe^{3+} + 2\,H_2O \tag{14}$$

This reaction (14) shows that acidic conditions are needed to ensure hydroxyl radical production. In fact, in acid solutions, the energy required to allow electron transfer from Fe^{2+} to H$_2$O$_2$ is minimal, and at the same time, peroxide decomposition due to the following reactions (15) and (16), is negligible:

$$H_2O_2 \rightarrow H^+ + HOO^- \tag{15}$$
$$H_2O_2 \rightarrow H_2O + 0.5\,O_2 \tag{16}$$

Conversely both these reactions are enhanced at neutral or basic conditions and in the presence of large amounts of Fe^{2+}.

Nevertheless extreme acid conditions may also reduce the oxidation effectiveness, because of the consumption of hydroxyl radicals through reaction (17):

$$HO^\bullet + H^+ + e^- \rightarrow H_2O \tag{17}$$

Application of Fenton's Treatment to Soil and Groundwater Remediation

The use of the Fenton's reagent has proven to be effective for the oxidation of a wide range of well known organic pollutants such as: BTEX (Lou and Lee, 1995), p-chlorophenol (Kwon et al. 1999), trihalomethanes (Tang and Tassos 1997), aromatic amines (Casero et al. 1997), ethylene glycol (McGinnis et al. 2001), nitrophenols (Goi and Trapido 2002), surfactants (Kitis 1999), pesticides (Scott et al. 1995; Pignatello et al. 1995), a few commercial azodyes and textile effluents (Kang and Chang, 1997; Szpyrkowicz et al. 2001; Kang et al. 2002) as well as wastewater and landfill leachates (Bae et al. 1997; Kang and Hwang 2000). On the contrary, Fenton's reaction has been rarely used for the treatment of inorganic pollutants like arsenic, in drinking water (Balarama Krishna et al. 2001).

It must be stressed that Fenton's reaction is the simplest way of producing hydroxyl radicals because no special reactants or special apparatus are required. Furthermore, iron and hydrogen peroxide are relatively safe chemicals (ITRC 2000). Consequently the whole system appears to be easy to handle, environmentally safe and cost effective.

The possibility of applying Fenton's process to contaminated soils was first demonstrated by Tyre et al. (1991), followed by Watts et al. (1993) in batch lab-scale experiments, by Ravikumar and Gurol (1994) with sand-packed column tests and by Kakarla and Watts (1997) with soil-packed column tests.

On the basis of these considerations, in many of the above cited studies, Fenton's reagent optimal conditions have been assessed in a quite restricted pH range, between 2 and 4, while at pH higher than pH=6, a very low effectiveness has been observed (Baciocchi et al. 2003). In order to achieve low pH conditions, soils may need to be acidified. The soil buffering capacity becomes then an important factor that must be considered in any feasibility study involving *in situ* Fenton's oxidation. Highly alkaline soils (containing limestone) may require a substantial acid addition to bring the pH into the optimal range, thus increasing the cost of the process. A pH adjustment may also be required during Fenton's treatment, due to the formation of H^+ ions from reaction (5) and (15).

In several works, the possibility of using the soil's naturally occurring iron as a catalyst, possibly without pH adjustment, in the so-called Fenton-like process, was also successfully investigated (Baciocchi et al. 2003).

In most of the experimental studies on the oxidation of organic compounds, the reaction times required to achieve a quite complete degradation were between 5 minutes and 2 hours (Arnold et al. 1995). However, reaction times from some hours even to some days were needed in case of complex substrates (Chen et al. 2001).

The efficiency of Fenton's oxidation is strongly affected by operative conditions (Kang and Hwang, 2000).

Fenton oxidation's kinetic follows generally a first order rate equation. However, there is a main factor which cause results to deviate from this trend. One of the limiting steps of the process is the regeneration of ferrous ions. As a consequence, Fenton's reaction development along time has been divided in two phases. First a quick substrate oxidation is generally observed, due to the initial rapid formation of hydroxyl radicals. The following decrease in the oxidation rate can be associated with the presence of large amounts of Fe^{3+} not converted to Fe^{2+}. The unfavorable Fe^{3+}/H_2O_2 ratio is then responsible for the slow oxidation of the remaining substrate.

There is a limit in the value of the Fe^{2+}/H_2O_2 ratio: the difficulty arises from the regeneration of ferrous ions. In effect, an accumulation of Fe^{3+} ions during the oxidation process is caused by reaction (1). This happens because the Fe^{2+} regeneration reactions (5) and (6) are slower. In addition, the Fe^{2+} concentration should be high enough to activate hydrogen peroxide's decomposition through reaction (1) but at the same time low enough to limit the undesirable reactions of Fe^{2+} (7) and (9).

As a result an optimal Fe^{2+}/H_2O_2 ratio has to be determined for each specific application. As an example, Arnold et al. (1995) found that Fenton oxidation of atrazine was greatly affected by the Fe^{2+}/H_2O_2 ratio. A ratio of 1:1 was found to be optimal, since increasing either the H_2O_2 concentration or the Fe^{2+} one, strongly stimulated the undesirable reactions.

Another parameter affecting the overall cost of the process is the H_2O_2 vs. contaminant ratio. A first order kinetic with respect to the concentration of Fenton's reagent , has been widely observed (Gallard and De Laat 2000). Consequently it would be expected that increasing the peroxide concentration would improve oxidation's effectiveness. However, excessive peroxide dosage can act as a scavenger of hydroxyl radicals according to reaction (8). It must be stressed that, according to reaction (1), an increase in peroxide concentration must always correspond to an increase in Fe^{2+} concentration. In wastewater treatment, the H_2O_2/contaminant ratio is typically maintained in the range between 1/1 and 5/1 (w/w). For soil treatment this ratio can increase up to 50/1, due to the oxidant demand of soil natural matter, contaminant hydrophobicity (expressed as: $logK_{OW}$, where K_{OW} is the octanol-water partition coefficient), and hydrogen peroxide instability (Quan et al. 2003). In some commercial applications of Fenton's reagent, a mixture of 5-50% wt. H_2O_2 was applied (Gates and Siegrist 1995; Cresap and Burke 2000; ITRC 2001). Some H_2O_2 mixtures contain therefore a large quantity of water. There is an advantage in applying peroxide at low concentrations, due to a resulting better control of heat and gas generation. This permits to reduce the resulting potential stripping of the contaminant. A concentration of 10% is generally recommended for several applications (ITRC 2001).

Experiments on MtBE Extraction by Soil Flushing and Fenton's Treatment of the Extracted Solution

In this section the results of an experimentation of MtBE extraction from a contaminated saturated soil, followed by an oxidation treatment of the extracted solution, are presented and discussed.

The experiments were performed on a soil artificially contaminated with MtBE. The soil was a sandy-loamy soil, whose composition is shown in Table 3 while selected chemical and physical characteristics are shown in Table 4.

Table 3. Composition of the soil used in soil flushing experiments

Component	Weight %	Particle size, mm
Silt	38.75	0-0.02
Clay	7.5	0.02-0.355
Sand	35	0.355-1.7
Gravel	18.75	1.7-9.5

Table 4. Characteristics of the soil used

pH	8.1
Organic carbon, %	3.2
Permeability, cm/s	$3.21 \cdot 10^{-3}$
Porosity, %	46
Humidity, g/kg	24.5
Bulk density, g/ml	1.30
Absolute density, g/ml	2.30

The soil was contaminated by flushing an aqueous solution of MtBE. The solution was flushed into the soil and continuously recirculated, until an homogeneous contamination and a residual level of MtBE in the saturated soil between 100 and 200 mg/l, were achieved.

The extraction tests were performed by flushing the soil with an aqueous solution of ethanol: two series of experimental tests were performed, at two different initial level of MtBE in the soil (80 and 100 mg/l). For each series of test, two flow rates of the extractant solution were tested (70 and 100 ml/h, corresponding to 3.4 and 4.8 m^3/m^2h respectively). The experiments were performed at the controlled temperature of 4°C, to avoid MtBE volatilization.

Several intervals of time were selected, each one corresponding to the flushing of a volume of solution equal to two times the value of the pore volume (PV) of the soil. The collected solution after extraction, was sampled and subjected to the analyses to determine MtBE and total organic carbon (TOC) content.

Figure 7 show the extraction yield observed in the four series of experimental tests.

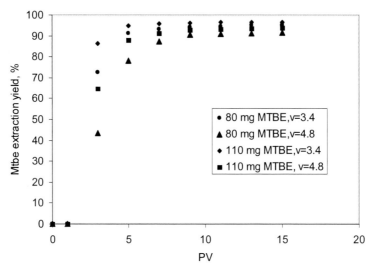

Figure 8. Experimental results of MtBE extraction by soil flushing.

As confirmed by Figure 8, an increase in MtBE extraction occurred when increasing volumes of solution were injected. At the same time, the extraction yield was higher when the initial amount of MtBE was higher and the contact time between the flushing solution and the soil was lower. A fast extraction was also observed: after the flushing of a volume of 5 PV of solution, an extraction yield of 91.1% and 94.8% was achieved for the initial concentration of 80 and 110 mg/l respectively.

The following table 5 provides the concentration of MtBE (C_e) and TOC in the extracted solution, together with the values of the flushed and recovered solutions' volumes (test with C_i=110 mg /200 g soil, v =70 ml/h).

The data reported in the above reported Figure 7 and Table 5, show that the relative amount of MtBE and the quantity of natural organic matter (NOM) extracted from the soil (measured as TOC), depended from the experimental conditions. Assuming that MtBE was not converted during the extraction, a mass balance was performed for each of the four experimental condition: the results are shown in Table 6.

Table 5. Experimental results of MtBE extraction by soil flushing
(C_i=110 mg /200 g soil, v =70 ml/h)

PV	Volume in (ml)	Volume out (ml)	TOC (mg/l)	C_e (mg/l)	Mtbe extraction yield (%)
1	69	/	/	/	/
3	207	128	743	741	86.2
5	345	260	93.0	72.0	94.8
7	483	398	42.2	5.58	95.6
9	621	536	31.0	4.14	96.0
11	759	670	20.2	2.17	96.3
13	897	808	12.5	0.82	96.4
15	1035	946	0.91	0.71	96.5

Table 6. Mass balances for the flushing experiments

MtBE initial amount (mg/200 g soil)	Flushing flow rate (ml/h)	MtBE extracted (mg)	NOM extracted (mg TOC)
80	70	76	68
80	100	72	56
110	70	106	50
110	100	103	34

MtBE extraction was not significantly dependent on the flushing solution's flow rate, while, this parameter greatly affected the extraction of organic matter from the soil due to its lower water solubility. In addition, higher amounts of MtBE lowered the quantity of NOM extracted. This could be explained considering the higher affinity of MtBE for the extraction agent.

The amount of MtBE volatilized could be considered negligible, due to the operating temperature of 4°C.

Mechanism of MtBE Oxidation with Fenton's Reagent

The solution recovered after the percolation of 5 PV of the flushing solution was then subjected to the oxidation treatment with Fenton's reagent. The initial MtBE concentration was 100 mg/l, and NOM concentration was in the range between 30 and 100 mg/l of TOC, depending upon the operating conditions.

The advantage of using chemical oxidation to treat a solution of water contaminated with MtBE, is that this compound has a very low biodegradability. Chemical oxidation can reduce the duration of the treatment and can eventually provide more biodegradable intermediates.

The radical attack on the MtBE molecule occurs on the side with less steric hindrance i.e. on the methyl group:

$$(CH_3)_3\text{-C-O-CH}_3 + HO^\circ \rightarrow H_2O + (CH_3)_3\text{-C-O-CH}_2^\circ$$

$$(CH_3)_3\text{-C-O-CH}_2^\circ + O_2 \rightarrow (CH_3)_3\text{-C-O-CH}_2\text{-O-O}^\circ$$

$$(CH_3)_3\text{-C-O-CH}_2\text{-O-O}^\circ + (CH_3)_3\text{-C-O-CH}_3 \rightarrow$$

$$\rightarrow (CH_3)_3\text{-C-O-CH}_2\text{-O-OH} + (CH_3)_3\text{-C-O-CH}_2^\circ$$

$$(CH_3)_3\text{-C-O-CH}_2\text{-O-OH} \rightarrow (CH_3)_3\text{-C-O-CHO} + H_2O$$

TBF

$$CO_2$$
$$\uparrow$$

Acid hydrolysis $(CH_3)_3\text{—C—O—CHO} \leftrightarrow (CH_3)_3C\text{—OH} + HCOOH$

                   TBF   TBA   formic acid

$$(CH_3)_3\text{—C—OH} \leftrightarrow (CH_3)_3\text{—C}^+ + H_2O \rightarrow CH_3\text{—C(CH}_2)\text{—CH}_3 + H^+$$

     TBA   Isobutene

$$CH_3\text{—C(CH}_2)\text{—CH}_3 + O_2 \rightarrow CH_3\text{—CO—CH}_3 + H\text{—CHO} \longrightarrow CO_2$$

     Isobutene   Acetone   formaldeide

This firstly produces TBF (tertiary-Butyl formate), an esther that successively is involved in a hydrolysis equilibrium because of the acidic environment. The hydrolysis products are TBA (tertiary-Butyl-alcohol) and formic acid. The latter one is then easily transformed in CO_2. TBA then can be oxidized and transformed in acetone, which represents the most abundant and persistent by-product. TBA can also be produced from a direct oxidation of TBF. Finally the oxidation process leads to the formation of methanol and, allowing longer reaction times, to the final conversion to CO_2.

Fenton Oxidation of the Extracted Solution

The experimental tests were performed in sealed vials, on samples of the flushing solution or of a reference solution prepared dissolving in tap water the same amount of MtBE extracted. This was done to study the effects of NOM on the process.

The solutions were acidified with HCl down to pH=3.5 (in the optimal range for Fenton treatment, as shown in the previous sections). Preliminary tests were carried out to evaluate the extent of volatilization in the operating conditions chosen for the oxidation treatment. Results of these tests showed that from a 100 mg/l aqueous solution of MtBE the 8% of the compound volatilized, while in the presence of 80 mg/l dissolved organic matter form soil this amount was reduced to about 3%.

After pH adjustment, the solution was oxidized using hydrogen peroxide coupled with Fe^{2+} (as $FeSO_4 \cdot 7H_2O$), under a constant agitation at ambient temperature. After two hours reaction time, the residual iron was precipitated and the liquid phase was then vacuum filtered and analyzed.

Table 7 shows the results obtained in three series of experimental tests, performed at three different Fe^{2+} concentration. The aim was to evaluate, for each test, the optimal H_2O_2/Fe^{2+} ratio. The concentrations of MtBE (after 5 min reaction time) and of all the main byproducts of its oxidation (as described in the previous section) are reported.

Table 7. Influence of reagent concentration on Fenton's oxidation of MtBE

Fe^{2+} (mM)	$H_2O_2/$ Fe^{2+}	MTBE (mg/l)	TBA (mg/l)	TBF (mg/l)	Methanol (mg/l)	Acetone (mg/l)
4.5	1	52.2	70.3	5.20	1.90	10.3
	1.6	10.2	61.0	10.8	13.1	20.7
	2	9.13	50.5	12.3	11.6	25.3
	3	11.5	52.6	12.9	ND	36.5
10	1.5	0.97	30	ND	9.30	40.7
	2	ND	14.8	ND	9.00	49.1
	2.5	1.20	18.5	10.2	10.6	53.0
	3	0.76	ND	18.3	8.60	44.7
	3.5	1.92	36.2	ND	14.4	50.7
	4	3.40	17.8	36.2	2.60	48.4
20	1	1.36	25.3	22	10.1	59.0
	2	6.64	22.9	46.4	11.5	51.0
	3	12.6	28.5	45.6	16.8	46.8

Results show that, as expected, there is a limit to the increase of the H_2O_2/Fe^{2+} ratio: the difficulty is in the regeneration of the ferrous ions. An accumulation of Fe^{3+} ions during the oxidation progress is in fact caused by reaction (1), that occurs faster than the Fe^{2+} regeneration reactions (5) and (6). Fe^{2+} concentration should be high enough to activate hydrogen peroxide but at the same time, low enough to limit undesirable reactions of Fe^{2+} (7) and (9).

In the investigated range the higher oxidation rate was observed when Fe^{2+} concentration was 10 mM and $H_2O_2/Fe^{2+} = 2$. For this ratio, a maximum MtBE conversion was in fact generally observed.

Since, as shown in the previous sections, Fenton reaction is strongly affected by pH, another sequence of tests was performed to evaluate the influence of the initial solution's pH on MtBE conversion. It must be stressed that, considering the great buffering capacity of the soil, the extraction of soil constituents such as carbonates and NOM, produced a non negligible acid consumption during the acidification prior to the addition of Fenton's reagents.

The results of these tests are shown in table 8, for an initial concentration of 190 mg/l MtBE and a NOM concentration of 30 mg/l.

Table 8. Influence of reaction pH on Fenton's oxidation of Mtbe

pH	MtBE oxidation yield (%)
2	99.0
3.5	99.3
4	99.2
5	99.0
7.5	97.2

MtBE oxidation yield was always very high, even at circum neutral pH conditions, and this is a promising aspect in the prospective of an application of this process, for the possibility to reduce acid consumption. As expected, the higher conversion rate was observed at an initial pH of 3.5, in the common optimal range for Fenton oxidation treatment.

To obtain some information about the reaction kinetics, Table 9 reports the results obtained after the first 2 minutes. Reactions were carried out at the previously determined optimal operating conditions ($[Fe^{2+}]$= 10 mM, H_2O_2/Fe^{2+}= 2, pH=3.5).

Results confirmed that the final product of MtBE conversion was acetone.

Table 9. Kinetic of MtBE degradation by Fenton reagent (pH=3.5)

Time (min)	MTBE (mg/l)	TBA (mg/l)	TBF (mg/l)	Methanol (mg/l)	Acetone (mg/l)
0	187.3	0	0	0	0
0.083	146.7	0	0	0	3.06
0.25	30.1	57.86	0	3.01	25.36
0.5	7.43	47.42	0	3.22	42.61
0.75	3.5	40.46	26.18	0.62	47
1	1.57	20.05	0	0.61	46.1
5	0	19.47	0	0	52.88

Finally, Table 10 shows the results of tests performed at the same conditions than Table 9, except for the initial pH of 5.5.

Table 10. Kinetic of MtBE degradation by Fenton reagent (pH=5.5)

Time (min)	MTBE (mg/l)	TBA (mg/l)	TBF (mg/l)	Methanol (mg/l)	Acetone (mg/l)
0	178.5	0	0	0	0
0.25	26.86	55.58	7.62	15.73	30.66
0.5	14.42	53.89	0	15.57	34.87
1	2.74	25.37	24.76	11.33	46.08
5	0.81	18.82	9.92	9.16	53.21

The results of the two tests are in accordance: in particular it can be observed that TBF was more stable in less acidic conditions.

In Situ Chemical Oxidation of MtBE in Saturated Media

Lab Scale Experiments

Figure 9 shows the results of preliminary experiments of MtBE chemical oxidation in water.

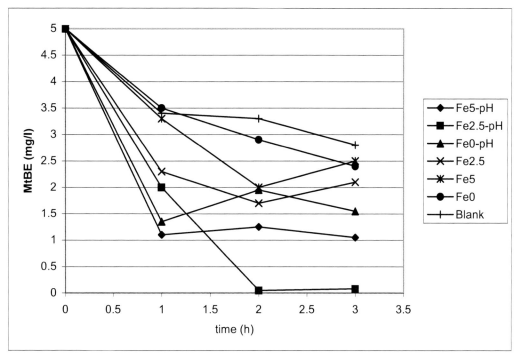

Figure 9. Batch removal tests in water.

The batch tests in water were performed in 250-ml beakers, filled with 50 ml of MtBE solution to achieve an initial concentration of 5 mg/l.

The theoretical amount of H_2O_2 required to oxidize MtBE was estimated on a stoichiometric basis (5.78 g of H_2O_2 for every 1.00 g of MtBE). Due to the presence in soil and groundwater of other contaminants and naturally occurring organic compounds in soil or groundwater, the quantity of oxidant required at field conditions is often five to seven time greater (Leethem, 2001). In these tests the quantity of H_2O_2 to be added in each beaker (19.48 g) chosen was five times greater than the stoichiometric quantity.

A quantity of 19.48 g of H_2O_2 was added in six of the seven beakers containing the MtBE solution. A parallel blank experiment was carried out adding distilled water instead the oxidant.

Table 11. Test conditions in the seven beakers of the batch removal tests in water

Test	Sample Name	$H_2O_2/MtBE$	$FeSO_4/H_2O_2$	pH
1	Blank	0	0	5,5
2	Fe0	5	0	5,5
3	Fe5	5	1/5	5,5
4	Fe2.5	5	1/2,5	5,5
5	Fe0-pH	5	0	3
6	Fe5-pH	5	1/5	3
7	Fe2.5-pH	5	1/2,5	3

Ferrous sulfate ($FeSO_4$) was chosen as a catalyst of Fenton's oxidation and added in four of the six beakers. Two different ratios of $FeSO_4$ and H_2O_2 solution (1/5 and 1/2,5) were compared to optimize the oxidation. In three of the six beakers pH was also reduced (adding sulfuric acid), to check if acidic condition could enhance the action of H_2O_2 in MtBE removal from water.

Comparing the results of the tests reported in Figure 9 the major evidences are:

- the pH significantly influences the effectiveness of MtBE oxidation;
- the action of removal is quite low in all of conditions tested;
- the highest removal efficiency occurred in the test Fe5 (38.1%), and in the test Fe2.5 when pH was adjusted thus confirming the importance of soil pH in Fenton's treatments.

Intermediate Scale Experiments

Following the lab scale tests, further experiments were performed at an intermediate scale, to assess the effectiveness of ISCO in MtBE removal from a saturated media.

As well as the above described tests with AS, the experimental set-up consisted of a steel tank (dimensions 1 x 1 x 1,2 m), containing 1 m^3 of soil saturated with an MtBE-contaminated solution. In this case the sparging point was substituted by the injection well represented in Figure 10, in order to allow the oxidant solution to spread within the media.

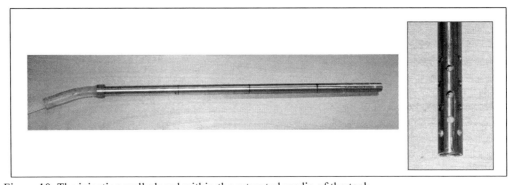

Figure 10. The injection well placed within the saturated media of the tank.

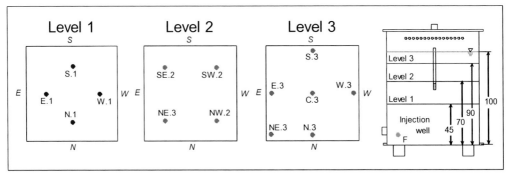

Figure 11. Injection well and sampling points location within the saturated porous media (all elevations are expressed in cm).

The position of the injection wells and the sampling points within the tank are reported in Figure 11.

The saturated media was contaminated with MtBE (6 mg/l). Ferrous sulfate ($FeSO_4$) was then added by mean of a peristaltic pump connected to the injection well and, finally, an H_2O_2 solution was injected. The oxidant concentration was based on MtBE reaction stoichiometry. In order to enhance the efficiency of treatment, a second injection of oxidant, without catalyst, was done 24 hours later than the first one.

Table 14 reports the concentration of MtBE along time in the different sampling points. The chemical analyses carried out on the solution samples also reveal the presence of acetone as the main by-product of the oxidation (not reported in Table 14), thus confirming the degradation pathway generally observed in Fenton oxidation of MtBE. The data of Table 14 have been plotted to better understand the trends of MtBE concentration measured within the contaminated media (Figure 12).

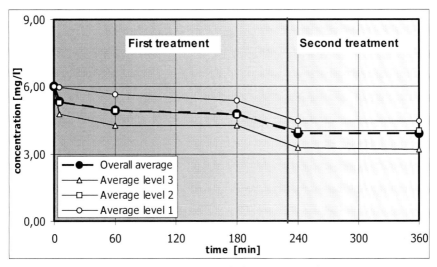

Figure 12. Trends of MtBE concentrations measured during the removal test.

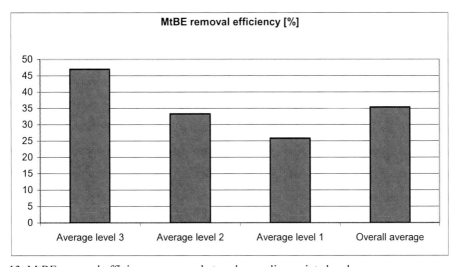

Figure 13. MtBE removal efficiency measured at each sampling points levels.

Table 14. Analytical results of MtBE concentrations measured at each sampling point during the removal test

Treatment	First			Second	
Time [min]	0	60	180	0	180
Sampling points	MtBE concentrations [mg/l]				
N III	6,81	5,48	5,50	2,77	2,64
E III	5,94	5,80	5,78	4,12	3,68
S III	6,78	5,52	5,53	4,36	4,25
NE III	4,35	4,57	4,50	5,00	5,43
C III	0,06	0,01	0,05	< 0,01	< 0,01
Average level 3	*4,79*	*4,28*	*4,27*	*3,25*	*3,20*
NW II	4,59	3,88	2,75	3,33	3,05
NE II	4,54	4,36	4,44	4,71	4,09
SW II	4,55	4,4	5,78	2,87	3,90
SE II	7,57	7,18	6,06	5,19	5,10
Average level 2	*5,31*	*4,96*	*4,76*	*4,03*	*4,04*
N I	5,18	4,31	4,27	4,14	4,05
E I	5,90	5,78	5,90	5,42	5,36
S I	6,27	6,30	6,22	4,24	4,16
W I	6,51	6,18	5,05	4,11	4,20
Average level 1	*5,97*	*5,64*	*5,36*	*4,48*	*4,44*
Overall average	5,36	4,96	4,80	3,92	3,89

The observed overall MtBE removal efficiency is presented in Figure 13.

The following conclusions can be drawn on the basis of the results of experiments.

- The results of the intermediate scale test are comparable with those obtained in the laboratory scale tests previous described.
- The overall average removal efficiency was 35,4%. However this data was highly dependent on the sampling level (47.0% in level 3, 33.3% in level 2 and 25.8% in level 1). The decreasing of the removal efficiency with the depth of the sampling levels, was probably due to the tendency of the oxidant solution to move upwards through the saturated media, maybe driven by the generation of O_2 in the solution during oxidation.
- Within the first hour of treatment the higher removal rate was observed, proving that oxidation is a fast reaction.
- Acetone resulted to be the main by-product of MtBE oxidation.
- Differences in MtBE concentrations measured among sampling points belonging to the same level were attributed to the presence of preferential pathways within the media. This indicates that the delivery of the oxidant solution through the porous media plays an important role in the overall removal efficiency of technology.
- As a consequence, the use of a more diluted oxidant solution may favour its distribution and, at the same time, may reduce heat generation thus limiting volatilization.

REFERENCES

Alimonti, C., Lausdei, D., Santoro L.B. (2006) Experimental analysis of the stripping process driven by air sparging systems, in: *Contaminated Soil, Sediments and Water: Successes and Challenges*, ed. E.J. Calabrese, P.T. Kostecki, J. Dragoon, vol. 10, 163-175, Springer.

Arnold, S. M., Hickey W. J. and Harris R. F. (1995) Degradation of atrazine by Fenton's reagent: condition optimisation and product quantification, *Environ. Sci. Technol.* 28, 2083-2088.

Baciocchi, R., Boni, M.L. and D'Aprile L. (2003) Hydrogen peroxide lifetime as an indicator of the efficiency of 3-chlorophenol Fenton's and Fenton-like oxidation in soils, *J. Hazard. Mater.* B96, 305-329.

Bae J., Kim S. and Chang H. (1997) Treatment of landfill leachates; ammonia removal via nitrification and denitrification and further COD reduction via Fenton's treatment followed by activated sludge, *Water Sci. Technol.* 36, 341-348.

Balarama Krishna M.V., Chandrasekaran K., Karunasagar D. and Arunachalam J. (2001) A combined treatment approach using Fenton's Reagent and zero valent iron for the removal of arsenic from drinking water, *J. Hazard. Mater.* B84, 229-240.

Beckett P.H.T., The use of extractants in studies on trace metals in soils, sewage sludges, and sludge-treated soils, Advances in soil science, 9 (1989), Springer-Verlag New York Inc.

Casero I., Sicilia D., Rubio S. and Pérez-Bendito D. (1997) Chemical Degradation of Aromatic Amines by Fenton's Reagent, *Wat. Res.* 31(8), 1985-1995.

Chang L.Y., A waste minimization study of a chelated copper complex in wastewater-treatability and process analysis, Waste management, 15 (1995) 209.

Chang M.C., Huang C.R., Shu H.Y. (2000) Effects of surfactants on extraction of phenanthrene in spiked sand, *Chemosphere* 41, 1295-1300.

Chen, G., Hoag, G.E., Chedda, P., Nadim, F., Woody, B.A. and Dobbs, G.M. (2001) The mechanism and applicability of *in situ* oxidation of trichloroethylene with Fenton's reagent, *J. Hazard. Mater.* B87, 171-186.

Cresap G.H. and Burke A.L. (2000) In situ chemical oxidation of fuel using hydrogen peroxide: a case study, *Hazard. Ind. Waste*, 32, 31-38.

Di Palma L., Ferrantelli P., Petrucci E. (2003) Experimental study of the remediation of atrazine contaminated soils through soil extraction and subsequent peroxidation, *J. Hazard. Mater.* 99(3), 265-276.

Di Palma, L. (2005) In situ chemical oxidation of contaminated environments with hazardous materials, in: *Soil and Sediment Remediation*, Ed. P. Lens, T. Grotenius, G. Malina and H. Tabuk, pagg. 200-222, IWA Publishing, London, UK, 2005.

Fenton H.J.H. (1894) Oxidation of Tartaric Acid in presence of Iron, *J. Chem. Soc.* 65, 899-910.

Freeman, F. (1975) Possible criteria for distinguishing between cyclic an acyclic activated complexes among cyclic activated complexes in addition reactions, *Chem. Rev.* 75, 439-491.

Freeman H. M. and Harris E. F. (1995) Hazardous waste remediation: innovative treatment technologies, Technomic Publishing Co. Inc., Lancaster, Basel.

Gallard and De Laat (2000) Kinetic modeling of Fe(III)/H2O2 oxidation reactions in dilute aqueous solution using atrazine as a model organic compound, *Wat. Res.*, 34, 3107-3116.

Gates D.D. and Siegrist R.L. (1995) *In situ* chemical oxidation of trichloroethylene using hydrogen peroxide, *J. Environ. Eng.*, 121(9), 639-644.

Ghassemi, M. (1988) Innovative in situ tratment technologiesb for cleanup of contaminated sites, *J. Hazard. Mater.* 17, 189-206.

Goi A. and Trapido M. (2002) Hydrogen peroxide photolysis, Fenton reagent and photo-Fenton for the degradation of nitrophenols: a comparative study, *Chemosphere* 46, 913-922.

Ho, C.L., Shebl, M.A.A. and Watts, R.J. (1995) Development of an injection system for in situ catalyzed peroxide remediation of contaminated soil, *Hazard. Wastes Hazard. Mater.* 12, 15-25.

ITRC (2000) Dense non aqueous pahase liquids (DNAPLs): Review of emerging characterization and remediation technologies, Interstate Technology and Regulatory Cooperation Work Group, USA.

ITRC (2001) Technical and regulatory guidance for *in situ* chemical oxidation of contaminated soil and groundwater, Interstate Technology and Regulatory Cooperation Work Group, USA.

Jafvert C.T. (1996) Surfactants/Cosolvents. Technology evaluation report, EPA Groundwater Remediation Technologies Analysis Center, Pittsburgh, PA.

Kakarla, P.K.C. and Watts, R.J. (1997) Depth of Fenton-like oxidation in remediation of surface soil, *J. Environ. Eng.* 123, 11.

Kang S.F. and Chang H.M. (1997) Coagulation of Textile Secondary Effluents with Fenton's Reagent, *Water Sci. Technol.* 36(12), 215-222.

Kang S.F., Liao C.H. and Chen M.C. (2002) Preoxidation and coagulation of textile wastewater by the Fenton process, *Chemosphere* 46(6), 923-928.

Kang Y.W. and Hwang K.Y. (2000) Effects of reaction conditions on the oxidation efficiency in the Fenton process, *Wat. Res.* 34(10), 2786-2790.

Kitis M., Adams C. D. and Daigger G.T. (1999) The effects of Fenton's Reagent pretreatment on the biodegradability of nonionic surfactants, *Wat. Res.* 33(11), 2561-2568.

Kuo W.G. (1992) Decolorizing dye wastewater with Fenton's reagent, *Wat. Res.* 26(7), 881-886.

Kwon B.G., Lee D.S., Kang N. and Yoon J. (1999) Characteristics of *p*-chlorophenol oxidation by Fenton's reagent, *Wat. Res.* 33(9), 2110-2118.

Lin S.H. and Lo C.C., (1997) Fenton Process for Treatment of Desizing Wastewater, *Wat. Res.* 31(8), 2050-2056.

Liu M., Roy D. and Wang G. (1995) Reactions and transport modeling of surfactants and anthracene in the soil flushing process, *Waste Manage.*, 15(5/6), 423-432.

Lou J.C. and Lee S.S. (1995) Chemical Oxidation of BTX Using Fenton's Reagent, *Hazard. Waste & Hazard. Mater.* 12(2), 185-193.

Lu M.C., Chen J.N. and Huang C.P. (1997) Effects of Inorganic Ions on the Oxidation of Dichlorvos Insecticide with Fenton's Reagent, *Chemosphere* 35(10), 2285-2295.

Maillard C., Guillard C. and Pichat P. (1992) Comparative Effects of the TiO_2-UV, H_2O_2-UV, H_2O_2-Fe^{++} Systems on the Disappearance Rate of the Benzamide and 4-hydroxibenzamide in Water, *Chemosphere* 24(8), 1085-1094.

Martin, S.B. and Allen, A.E., Recycling EDTA after heavy metals extraction. Chemtech, 26 (1996) 23-25.

McGinnis B.D., Adams V.D. and Middlebrooks E. J. (2001) Degradation of ethylene glycol using Fenton's reagent and UV, *Chemosphere* 45(1), 101-108.

Mohanty N.R. and Wei I.W. (1993) Oxidation of 2,4-Dinitrotoluene Using Fenton's Reagent: Reaction Mechanisms and Their Practical Applications, *Hazard. Waste Hazard. Mater.* 10(2), 171-183.

Pearlman, R.S., Yalkowsky, S.H. and Banerjee, S. (1984) Water solubilities of polynuclear aromatic and heteroaromatic compounds, *J. Phys. Chem. Ref. Data* 13(2), 555-62.

Pignatello J.J., Sun Y. (1995) Complete Oxidation of Metolachlor and Methyl Parathion in Water by The Photoassisted Fenton Reaction, *Wat. Res.* 29(8), 1837-1844.

Quan, H.N., Teel, A.L. and Watts, R.J. (2003) Effect of contaminant hydrophobicity on hydrogen peroxide dosage requirements in the Fenton-like treatment of soils, *J. Hazard. Mater.* B102, 277-289.

Ravikumar, J.X. and Gurol M.D. (1994) Chemical oxidation of chlorinated organics by hydrogen peroxide in the presence of sand, *Environ. Sci. Technol.* 28(3), 394-400.

Roote D.S. (1998) Technology status report: *In situ* soil flushing, EPA Groundwater Remediation Technologies Analysis Center, Pittsburgh, PA.

Scott A.M., Hickey W.J. and Harris R.F. (1995) Degradation of Atrazine by Fenton's reagent: condition optimization and product quantification, *Environ. Sci. Technol.* 29, 2083-2089.

Siegrist R.L. (1998) *In situ* chemical oxidation: Technology features and applications, *Conference on Advances in Innovative Ground-water Remediation Technologies, Atlanta, GA, 15 December 1998*, Ground-water Remediation Technology Analysis Center, US EPA Technology Innovative Office.

Sun, B., Zhao, F.J., Lombi E. and McGrath S.P., Leaching of heavy metals from contaminated soils using EDTA. Environmental Pollution, 113 (2001) 111-120.

Szpyrkowicz L., Iuzzolino C. and Kaul S.N. (2001) A Comparative Study on Oxidation of Disperse Dyes by Electrochemical Process, Ozone, Hypochlorite and Fenton Reagent. *Wat. Res.* 35(9), 2129-2136.

Tang W.Z. and Tassos S. (1997) Oxidation Kinetics and Mechanisms of Trihalometanes by Fenton's Reagent. *Wat. Res.* 31(5), 1117-1125.

Tarr M.A. and Lindsay M.E. (1998) Role of dissolved organic matter in Fenton degradation of hydrophobic pollutants, Abstract presented at the 19th Annual Meeting of SETAC, Nov. 15-19, Charlotte, NC, USA.

Tejowulan, R.S., Removal of trace metals from contaminated soils using EDTA incorporating resin trapping techniques, Environmental Pollution, 103 (1998) 135.

Tyre B.W., Watts R.J. and Miller G.C. (1991) Treatment of four biorefractory contaminants in soil using catalyzed hydrogen peroxide, *J. Envron. Qual.*, 20(4), 832-838.

US EPA - Environmental Protection Agency (1998) Field applications of *in situ* remediation technologies: Chemical Oxidation, EPA 542-R-98-008.

US EPA - Environmental Protection Agency (2002) Arsenic treatment technologies for soil, waste and water, *Solid Waste and Emergency Response* EPA 542-R-02-004.

Venkatadri R. and Peters R.W. (1993) Chemical Oxidation Technologies: Ultraviolet Light/Hydrogen Peroxide, Fenton's Reagent and Titanium Dioxide-Assisted Photocatalysis, *Hazard. Waste & Hazard. Mater.* 10(2), 107-149.

Yin Y. and Allen H.E. (1999) *In situ* chemical treatment, Groundwater Remediation Technologies Analysis Center, Technology Evaluation Report, TE-99-01, Pittsburgh, PA, 82 pp.

Walling C. (1975) Fenton's Reagent Rivisited, *Acc. Chem. Research* 8, 125-131.

Watts, R.J., Udell, M.D. and Mansen, R.M. (1993) Use of iron minerals in optimizing the peroxide treatment of contaminated soils, *Water Environ. Res.* 65, 839-844.

Weiss J. (1935) The Catalytic Decomposition of Hydrogen Peroxide on different Metals. *Trans. Farad. Soc.* 3L, 1547-1557.

West, C.C. and Harwell, J.H. (1992) Surfactants and subsurface remediation, *Environ. Sci. Technol.* 26(12), 2324-2329.

Zepp R.G., Faust B.C. and Hoigné J. (1992) Hydroxyl Radical Formation in Aqueous Reactions (pH 3-8) of Iron(II) with Hydrogen Peroxide: the Photo-Fenton Reaction. *Environ. Sci. Technol.* 26, 313-319.

In: Groundwater Research and Issues   ISBN: 978-1-60456-230-9
Editors: W. B. Porter, C. E. Bennington, pp. 29-44   © 2008 Nova Science Publishers, Inc.

Chapter 2

SUSTAINABLE RATES OF GROUNDWATER EXTRACTION IN AN URBANIZED PLAIN, CENTRAL JAPAN

Adrian H. Gallardo[*1], *Atsunao Marui*[1], *Shinji Takeda*[2] *and Fumio Okuda*[2]

[1]AIST, Geological Survey of Japan、 Higashi 1-1-1, Central 7,
Tsukuba 305-8567, Japan
[2]Hytec Co. Yodogawa-ku, Miyahara 2-11-9,
Osaka 532-0003, Japan

ABSTRACT

Land subsidence due to excessive extraction of groundwater is a major environmental problem that has affected for several decades both the urban centers and the surrounding peri-urban farmlands scattered within the Nobi Plain, Central Japan. Even though strict regulations successfully reduced the subsidence rates, the threat is still far from being eliminated. The plain is one of the most important economic centers of Japan, sustaining an intensive agricultural and industrial production. Furthermore, its population density is among the highest in the country, averaging more than 1,000 persons per km^2. The recent economical growth of the region is expected to be accompanied by new developments and a higher demand for water supply; therefore, authorities are compelled to update management strategies to satisfy the increasing demand without causing undesirable effects on the environment. Knowledge of the groundwater availability becomes an essential requisite for the establishment of these policies. A numerical simulation was applied to understand the groundwater flow in the region. The resulting model was then used to determine the safe yield of the aquifers under different management options. The analysis concentrated on the Aburashima site, as this area faces the possibility of development in the forthcoming years. Calculations

* Corresponding author: ad.gallardo@aist.go.jp

indicated that withdrawal from the shallow aquifer is limited. In addition, a decline in piezometric levels reduced the groundwater availability in summer. Spreading a number of wells around the site constitutes an alternative to increase the pumping volume. Higher quantities and lower extraction costs makes the lower aquifer a more reliable source of groundwater. The present study constitutes one of the first steps towards the implementation of an integrated management plan of the water resources in the Nobi Plain, and constitutes a valid reference to update the current legislation regarding groundwater use in the region.

1. INTRODUCTION

With an area of about 1800 km^2 and approximately 4.5 million people, the Nobi Plain is the second largest populated region in Japan, only behind the Tokyo metropolis and surroundings. As in many other urban areas in Asia (e.g., Tokyo, Niigata, Osaka, Shangai, Bangkok, Hanoi, Calcutta, Taipei), excessive groundwater abstraction from a series of confined aquifers resulted in compaction of clay interbeds and the consequent land subsidence of the region. The issue dates back to the 1950's when the area underwent an explosive development following the post-war recovery. The subsidence increased exponentially and registered a peak in 1973 when the elevation at some areas decreased more than 20 cm (Yamamoto, 1984). Land subsidence can have several negative implications, as damage to buildings and roads, groundwater quality deterioration and salt encroachment, decline in storage capacity, change in flow patterns, and localized flooding (Larson et al. 2001). Given the magnitude of the problem, a number of regulations on groundwater pumping were enforced from 1974, effectively bringing the situation under control. As a result, areas subsiding more than 1 cm/yr reduced from 293 km^2 in 1975 to 9 km^2 in 2004 (METI, 2006).

Despite the success of the implemented strategies, the recent economic growth of the region is promoting new industrial and agricultural developments, which are accompanied by an increase in the demand of groundwater. Thus, the management authorities are compelled to meet the consumer requirements while preventing undesirable effects on such a sensitive environment. Aware of these needs, the prefecture of Gifu supported the present work towards the establishment of a modern plan for management of the water resources in the Nobi Plain. Results constitute also a scientific tool useful to update the current legislation on groundwater exploitation in the region. The analysis concentrated on the Aburashima site, southwest of the Plain, as it locates at the tripartite boundary of the Aichi, Gifu and Mie prefectures, all of them involved in the implementation of control policies within the basin. In addition, the site faces the possibility of new projects in the forthcoming years which make it especially susceptible to unacceptable consequences. Understanding the groundwater movement will ensure its proper utilization (Don et al. 2005) therefore, a numerical simulation was applied to obtain a detailed picture of the flow patterns over the region. After calibration, the model was utilized to determine the maximum rates of groundwater feasible to be extracted on a sustainable basis at Aburashima without causing a negative impact into the environment. Finally, some alternatives were evaluated to optimize the water supply from the aquifers.

2. OVERVIEW OF THE AREA

The Nobi Plain occupies an area of about 45 km by side located within the prefectures of Gifu, Aichi and Mie, in the central part of Japan. It is surrounded by the Paleozoic basement of the Owari Hills to the north and east, the Yoro Mountains to the west, and the Ise Bay on the south (Figure 1). Development of Nagoya and neighbor cities transformed the area into a major industrial center with several millions of residents. Four large rivers (Ibi, Nagara, Iso and Shounai) and a number of minor tributaries characterize the surface waters of the region. Despite the existence of an adequate channeling and tap system, there is still a heavy dependence on groundwater for public supply. The plain lies in a tectonic basin tilted westward, and filled with variable sediments mainly deposited during the Pliocene and Pleistocene (Figure 2). The main aquifers comprise three layers of river-bed gravel deposited during glacial advances (Uchida et al. 2003). From surface to depth, these aquifers are called the Fist, Second, and Third Gravel Bed, also known as the G-1, G-2, and G-3 respectively. Groundwater is mainly withdrawn from the shallow unit, while water for industrial use is mostly pumped up from the Second Gravel Bed (Yoshida et al. 1991). Since most production wells concentrate in the first and second aquifer, the present study restricted to a depth of 200 m, in coincidence with the base of the G-2. As already stated, the large amount of pumping has caused huge problems of subsidence and salinization, and as result, a large part of the Nobi Plain is below sea level (Yamanaka and Kumagai, 2006).

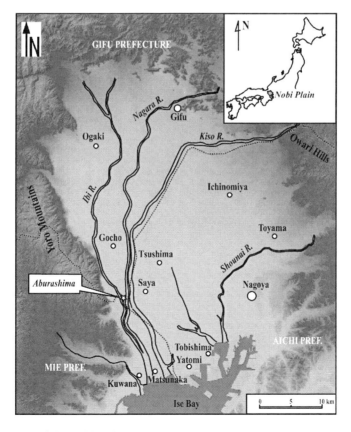

Figure 1. Location map of the Nobi Plain.

Period	Form	ation	Remarks
	Holocene N	anyo	Alluvium. Transgression
Quaternary	Pleistocene	Nobi	Last Glacial
		First Gravel Bed	
		Atsuta Last	Interglacial
		Second Gravel Bed	Transgression and Regression
		Ama	
		Third Gravel Bed	
		Yatomi	
Neogene	Pliocene	Tokai Group	Tokai Lake

Figure 2. Summary of stratigraphy and main geologic events in the Nobi Plain district (after Sakamoto et al. 1986).

The aquifers in consideration are separated by a series of silt and clays which limit the vertical movement of groundwater and promote the confining conditions of the aquifers. At the top of the sequence, the Nanyo and the Nobi formations are formed by a thin layer of alluvial sands followed by nearly 40 m of clays with a hydraulic conductivity in the order of 10^{-8} cm/sec. On the other hand, the First and the Second aquifers are separated by the Atsuta formation, deposited in the inner bay and its surroundings during the Last Interglacial and early half of the Last Glacial Times (Sakamoto et al. 1986). Its upper member is composed by 33 m of fine to medium sands intercalated with thin horizons of clay rich in organic matter. In contrast, the lower end is formed almost exclusively by densely packed clays reaching a thickness of about 40 m at the investigation site. Hydraulic conductivity ranges from 2 x 10^{-4} cm/sec at the top of the formation, to 4 x 10^{-8} within the deeper clays (Table 1).

The Japan Meteorological Agency maintains updated records of precipitation throughout the country. Data for the period 1986-2005 was used to calculate the mean rainfall within the plain. The highest rates corresponded to the Ogaki station in the north, with an average of 1876 mm/yr. Quantities reduced to 1676 mm/yr at Aichi station in the center of the plain, and reached a minimum of 1596 mm/yr at Kuwana city, near the Ise Bay. There is no information about evapotranspiration so, rates were calculated by the Thornthwaite method (1948). The procedure is based on the assumption that potential evapotranspiration is dependent only upon meteorological conditions, ignoring the effects of vegetation. While this assumption is not exact, the method is reasonably accurate in determining annual values, especially in humid areas (Fetter, 2000). The potential evapotranspiration (ET) is then calculated by the formula:

$$ET \ (cm/month) = 16(\frac{10T}{I})^{a} \qquad (1)$$

where T is the monthly temperature (°C) and I is the annual heat index defined as

$$I = \sum_{i=1}^{12} (Ti/5)^{1.514} \quad i: 1\sim12 \qquad\qquad (2)$$

$$a = (492390 + 17920\,I - 77.1\,I^2 + 0.675\,I^3) \times 10^{-6}$$

Considering a maximum temperature of 26.5 °C, the calculated evapotranspiration was 774 mm/yr in Ogaki, 767 in Kuwana, and 747 in the Aichi station. These values indicate that 41 to 48 % of the rainfall is lost to the atmosphere.

Table 1. Physical characteristics of the geological system

Formation		Main lithology	Hydraulic conductivity (cm/sec)	Thickness* (m)
Nanyo		Fine sands - silt -	3.2×10^{-3} to 9.9×10^{-8}	36.4
Nobi		clay	4.2×10^{-4} to 7.8×10^{-8}	20.7
First Gravel Bed		Clays - silt Gravel - sands	2.1×10^{-3} to 1.1×10^{-4}	15.8
Atsuta	upper	Fine sands	1.9×10^{-3} to 6.9×10^{-7}	60
	lower	Sands - clays	2.1×10^{-3} to 4.4×10^{-8}	42.1
Second Gravel Bed		Gravel - sands	2×10^{-3} to 3.3×10^{-3}	25

* measured at Aburashima.

3. METHODS OF STUDY

Groundwater modeling is a powerful management tool which can serve multiple purposes such as allowing quantitative prediction of the responses of the hydrologic system to externally applied stresses (Senthilkumar and Elango, 2004). Thus, after gathering the hydrogeological data of the area a three-dimensional model was constructed to understand the groundwater flow in the Nobi Plain. The model was refined around Aburashima to account for local details. Based on the flow simulation, a series of scenarios were analyzed to estimate the maximum amount of groundwater that can be extracted at the site of interest without leading to land subsidence.

The investigation started by drilling three exploratory wells to a maximum depth of 200 m. These wells permitted define the geometry and physical properties of the sediments in the site, which constituted a basic input to the model.

The next phase of the research was based on numerical modeling using the modular three-dimensional finite-difference code Modflow (McDonald and Harbaugh, 1988). This program is widely applied for simulating flow in complicated basins with various natural and/or artificial hydrogeological processes (Ebraheem et al. 2004). Modeling of the groundwater flow through porous media is performed by solving the governing equation:

$$\frac{\partial}{\partial x}(Kxx\frac{\partial h}{\partial x}) + \frac{\partial}{\partial y}(Kyy\frac{\partial h}{\partial y}) + \frac{\partial}{\partial z}(Kzz\frac{\partial h}{\partial z}) - W = \text{Ss}\frac{h}{\partial t} \quad (3)$$

where *Kxx*, *Kyy* and *Kzz* represents the saturated hydraulic conductivity along the x, y and z directions; *h* is the potentiometric head above a common datum; W is a volumetric flux per unit volume and represents sources and/or sinks of water; *Ss* is the specific storage of the porous medium, and *t* corresponds to time. Since the simulation was run under steady state, the terms on the right were neglected. The model domain was replaced by a series of discrete points in space, and hydraulic heads at each one of them was obtained solving the governing equation. After development, the flow model was calibrated to the site conditions to ensure it provided a reasonable representation of the real situation. On the basis of the computed results, the simulation was finally used to determine the safe yield of the aquifers at Aburashima.

4. NUMERICAL MODELING

4.1. Concept and Data Input

The model domain was delimited by the Yoro Mountains and Owari Hills to the west, north, and northeast, the Ise Bay to the south, and the Shounai River to the southeast. Thus, a large part of the region was circumscribed by impermeable boundaries, while a constant head of 0 m was assigned to the bay. River cells were imposed along the major bodies of surface water. For management purposes the city of Nagoya constitutes an independent entity therefore, it had to be excluded from the analysis. The geology of the area was divided into seven layers, two of them representing the upper and lower aquifer, the rest attributed to the different confining units. Each layer was assumed to be continuous, with a general tilt westwards. The surface elevation was set in accordance with the digital maps edited by the Geographical Survey of Japan (1997).

The model grid totalized 314 rows and 198 columns. Cells were of variable size although they were refined to a maximum of 20 m by side around Aburashima to get better representation of the piezometric contours (Figure 3). Hydraulic conductivity was assigned to each individual layer, with values ranging from 2×10^{-3} in the sands and gravels, to 9×10^{-8} cm/sec in the clay horizons.

Aside from rainfall, recharge to the aquifers is dependant on variables such as land use, soil type and moisture, evapotranspiration rates, irrigation, topography, and period of the year. A detailed evaluation of all these contributions is highly complex so, the overall recharge to the upper layer was initially approached from the difference between rainfall and evapotranspiration, and then reasonably adjusted during the calibration process to reproduce the observed site-conditions.

Pumping was represented by negative fluxes which incorporated the groundwater withdrawal in 822 wells at the First Gravel Bed, and 737 wells at the lower aquifer. Pumping wells in Gifu prefecture were grouped in coincidence with the main industrial centers, while for Aichi, they were assigned in accordance with a 1 km x 1 km mesh maintained by the prefecture. Extraction rates were represented as the annual mean discharge for each well during 2004, coincident with the most recent information available.

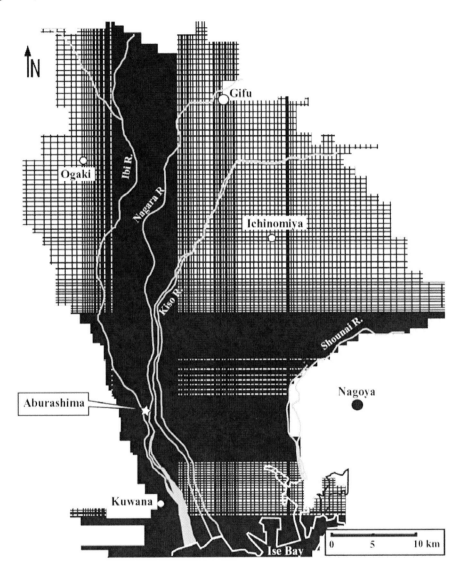

Figure 3. Model domain and finite-difference grid.

4.2. Model Calibration

The steady-state model was calibrated against in-situ data until simulated results reasonably matched the conditions observed in a network of 69 monitoring wells, providing a good coverage of the Plain. In agreement with the groundwater abstraction data, calibration was based on potentiometric observations for 2004. The procedure consisted in systematically varying the poorly known values of recharge and river stages, without introducing any changes to better defined parameters. When results were not satisfactory, the input values were modified and a new calibration process was started. In a second step, the optimization process was restricted specifically to the site of interest until no significant improvements in the results were obtained. The best outcomes were achieved after dividing the model domain in 11 regions of recharge with rates from 0 to 900 mm/yr. Maximum recharge would take place on the northwest along the boundary with the Yoro Mountains, while values can be considered minimum towards the lowlands in the center and south. Water stage and the conductance of rivers and streams were also adjusted to produce more realistic results. Figure 4 shows the difference between monitored and computed heads after calibration. The mean residual for the model is 0.048 m, which represents 0.15 % of the total head difference in the entire region (19.3 to -12.8 m). The absolute residual reached a minimum of 0.012 at the Tsushima monitoring station, but on the contrary, major bias led to poor results at Ichinomiya and surroundings, with a difference in heads up to 4.7 m. A misrepresentation of the realistic situation in that zone may be attributed to an insufficient grid refinement. In effect, the elevation of the water table varies drastically over short distances resulting thus in an especially high hydraulic gradient which can barely be represented by the selected mesh size, approximately 500 m by side in the area. Unfortunately, a further mesh refinement was not possible as it resulted in numerical instability and therefore, limited the capacity of the model to accurately simulate the groundwater heads at specific locations.

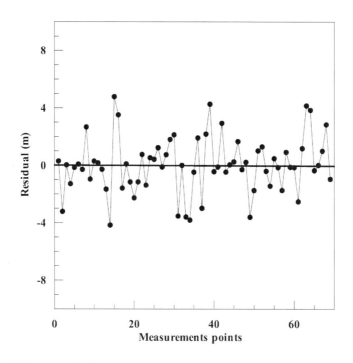

Figure 4. Residual of computed and observed heads at the monitoring wells.

Statistics parameters were also used to evaluate the accuracy of the calibration. The normalized root mean squared is the root mean squared error divided by the maximum difference in the observed head values. A reasonable calibration can be expected when it is below 10 % (Waterloo Hydrogeologic, 2005). The present simulation yielded a value of 6.7 %. In addition, a histogram of the residuals indicates that most of the residuals followed a normal distribution with the mean value near zero. As stated above, deviations in the distributions corresponded to the Ichinomiya surroundings (Figure 5).

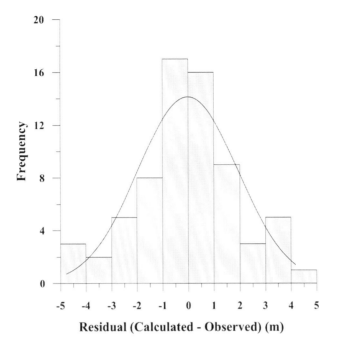

Figure 5. Distribution of residuals throughout the Nobi Plain.

Finally, an overall mass balance of the region yielded negligible differences between inflows and outflows constituting thus another indicator of a successful simulation. In this regard, approximately 48 % of the total water input came from infiltration through the riverbeds, while rainfall contributed with 37 % of the recharge and the constant heads cells 15 %. In contrast, the outputs corresponded to wells abstraction (52 %), discharge through constant heads (24 %), flow to rivers (16 %), and a minor part to evapotranspiration (8 %). A sensitivity analysis was also carried out to test the response of the model to variations in the key input parameters. It was found that the model is especially sensitive to changes in river conditions, as simulated heads for the upper aquifer decreased linearly with a reduction in the river stages. These results emphasize that river information must be carefully managed, as an error in the estimations translates into deviations of the same magnitude in the heads at the site of interest.

4.3. Flow Conditions

Head contours indicate that groundwater flows mostly from northwest to southeast. There is an additional component that originates at the foot of the Yoro Mountains and moves from west to east. Groundwater flow converges from all sides towards the center of the plain discharging ultimately through the Shounai River (Figure 6). There are many pumping wells used for industry in that part of the plain, which causes a low hydraulic head (Uchida et al. 2003). This statement was confirmed by the negative hydraulic heads simulated all over the central region.

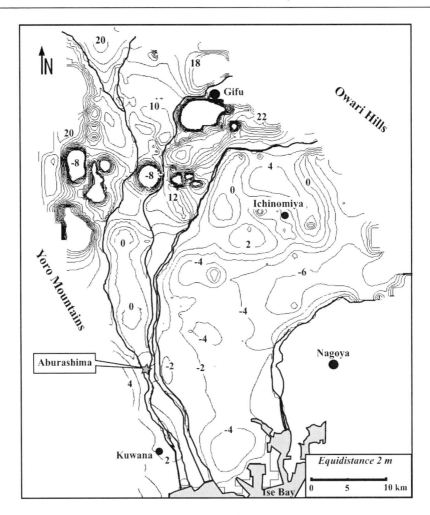

Figure 6. Simulated hydraulic heads for the shallow aquifer.

It is widely known that prolonged overdraft of groundwater can lead to salt contamination of freshwater aquifers. According to the simulation, groundwater abstraction resulted in the advance of seawater to a distance of about 2.5 km inland through the shallow aquifer, defining a mixing zone that encompasses the towns of Kuwana, Matsunaka, Yatomi and Tobishima near the coast. Exploitation of the deep aquifer is more limited and consequently, inland penetration of the saline wedge seems not to occur as a homogeneous front, but rather as a reduced tongue restricted to the surroundings of Yatomi and Tobishima. Chemical samples of the aquifers revealed that waters rich in Na-K and Cl⁻ dominate throughout the south of the plain spreading several km inland from the shoreline (Figure 7). Chloride concentrations as high as 3.5 g/L have been occasionally measured in the groundwater of Aburashima (METI, 2007). Unlike the sharp transition zone between fresh and brackish groundwater near the coastline depicted by most hydrogeology textbooks, saline groundwater was found far inland as occur in other coastal plains like in The Netherlands, Suriname, and Java (Groen et al. 2000). Although groundwater exploitation is important, the current extent of salinity should be interpreted carefully because it can be influenced by natural processes in addition to seawater intrusion. Chloride concentrations seems to increase together with a decrease in the sediments grain-size (METI, 2007), suggesting a possible contribution of fossil waters flushed from the marine clays and silts intercalated within the confining units. This hypothesis must be tested by further geochemical surveys supported by isotopic tracers, which constitute one of the most useful tools to discriminate Na-Cl sources in groundwater. Regardless the salinity origin, the hydrochemical zonation of the aquifers is a critical aspect to consider when planning future developments in the area, as groundwater of poor quality may be found in the new boreholes. Thus, utilization of the resource may become problematic for certain uses as agriculture, since high salinity of waters severely affects plants growth and reduces yields.

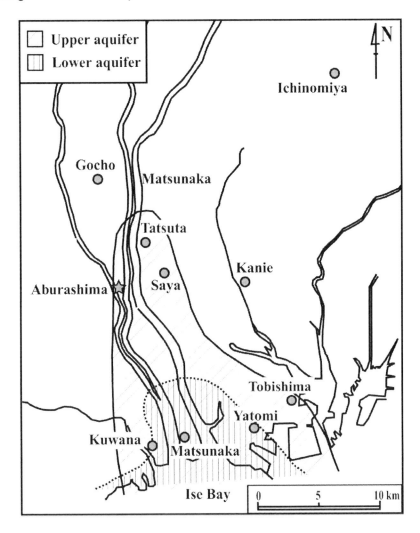

Figure 7. Spatial extent of saline waters (after METI, 2006).

5. Predictions for Aquifers Exploitation

5.1. Groundwater Availability

The validated model was ultimately used to calculate the availability of groundwater with the aim at maximizing the pumping extraction from the confined aquifers while preventing subsidence. Long-term monitoring at Gōcho station showed the site is especially susceptible to land subsidence when the groundwater of the upper aquifer drops more than 0.8 m below the average level (METI, 2007). Given its proximity, it is expected the system at Aburashima will respond in a similar manner to head fluctuations. Since there was a difference of 0.17 m between model results and observations, predictions at Aburashima conservatively considered the permissible groundwater drawdown cannot exceed 0.63 m. In contrast, data at the Tsushima and Saya stations showed that during 1985-1986, a reduction in pumping from the Second aquifer was accompanied by a rebound of groundwater levels above -10 m, giving place at the same time to the stabilization of subsidence rates to a minimum. From the above information, it can be concluded that sustainable pumping in the Aburashima site can only be achieved if the maximum drawdown does not exceed 0.63 m for the upper aquifer, and 10 m for the deep unit.

It was calculated that the maximum supply capacity of one well at the shallow aquifer is 27.7 m^3/day, while the deep one permits to extract 28 times more water, with a maximum rate of 776.4 m^3/day. A possible strategy to increase the extraction volumes consists in placing a number of wells in a circle centered in Aburashima. As can be anticipated, raising the number of wells is corresponded with an increase in the pumping rates (Figure 8). However, the relation is not nonlinear, especially for the deep aquifer, which means the extraction efficiency reduces for larger rates. Generally speaking, constrains in the exploitation efficiency translate directly into a higher cost for water supply. Then, it is possible to estimate the best well design from the economical point of view. Considering the site location, sediments characteristics, and the current market conditions in Japan, the cost for constructing an extraction well to the upper aquifer (60 m) is about U$$ 40,000. The price approximately duplicates when drilling 175 m to the Second Gravel Bed (Hytec Co., personal communication). Then, it was found that for the shallow aquifer, the best compromise between exploitation rates and cost is attained for 5 wells, where the price of water reaches a minimum (Table 2). As shown, at the early stages the price is about 44.5 $/m^3, but it decreases exponentially to 3.7 $/m^3 after 1 year of pumping, as the investment is better amortized over longer consumption periods. The cost of water delivery is also a function of the vertical distance, as deeper aquifers require a lift over greater distances and against a higher gravity force. In the study site, the benefit of extracting much larger volumes of groundwater from the lower aquifer overcomes the higher costs of lifting, resulting in a more efficient relationship extraction-price.

As an example, a $720,000 investment will allow the construction of 18 wells into the first aquifer and only 9 into the deep one; however, it is more effective to extract water from the lower unit as there are both a 6.7 times increase in production and a simultaneous drop in the prices with respect to the shallow aquifer. It must be emphasized that the analysis above is far from being a study of costs but instead, it must be treated as a complementary exercise to help finding an optimal solution in terms of total pumping rate and consumer benefits. The water price was calculated on the cost of wells and pumping volumes alone; therefore, it reflects the bottom-end value expected as other factors like wells maintenance, energy requirements, and operating conditions will add up to the true extraction expenses. Furthermore, the prices are sensitive to the period of production, decreasing exponentially especially after 6 months of withdrawal.

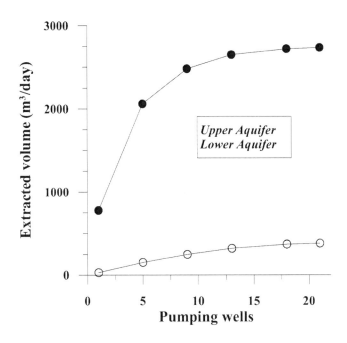

Figure 8. Maximum pumping rates in relation to the number of wells.

Table 2. Cost of groundwater extraction as a function of the number of wells and period of pumping. Unit: US$/m^3

No. wells	G-1				G-2			
	30 days	90 days	180 days	1 year	30 days	90 days	180 days	1 year
1	48.1	16.0	8.0	4.0	3.4	1.1	0.6	0.4
5	44.5	14.8	7.4	3.7	6.5	2.2	1.1	0.7
9	48.6	16.2	8.1	4.0	9.7	3.2	1.6	1.1
13	54.2	18.1	9.0	4.5	13.1	4.4	2.2	1.5
18	65.0	21.7	10.8	5.3	17.7	5.9	2.9	2.0
21	73.5	24.5	12.3	6.0	20.5	6.8	3.4	2.3

5.2. Safe Yield

Calculations were done under steady state to determine the mean groundwater available for exploitation. Nevertheless, intensive withdrawal for irrigation of rice fields in summer causes some temporal declines in the groundwater levels that reduce the supply capacity of the aquifers. Data collected during the second half of 2006 indicate that groundwater heads in Aburashima declined up to 0.24 m respect the values used for the model predictions. This drop in heads is negligible for the lower aquifer with an allowable drawdown of 10 m however, it becomes highly significant for the shallow unit, where the permissible drawdown is in the order of cm. Low groundwater-levels may last from a few days to several months therefore, the extraction policies for the First Gravel Bed should be defined on a seasonal basis, otherwise determined in relation to the deepest heads observed in the field. With a water table falling to a minimum, the permissible drawdown of the upper aquifer reduces and in consequence, the extractable flow rates decreases between 38.5 and 44.5 %, nearly 8.5 to 12 m^3/day per production well. These results show the strong influence that agricultural activities exert on the supply capacity of the shallow aquifer so, they must be carefully considered in the establishment of new abstraction policies. Furthermore, changes in groundwater recharge due to climatic variations as droughts, fluctuations of the stream flow in rivers, and the effect of local heterogeneities may result in incorrect determinations of the limit of pumpage from the aquifer. Thus, an effective management of the water resources must consider all components of the hydrologic system to maintain a sustainable yield in the long term (Sophocleus, 1988; Sakiyan and Yazicigil, 2004). Therefore, a "safety factor" of 20 % was added to the supply capacity of the aquifer when heads are at their lowest, so as to ensure that pumping will not lead to subsidence even during periods of reduced groundwater availability. These rates constitute the safe yield of the aquifer, and can be regarded as the maximum amount of groundwater available for stakeholders in Aburashima (Table 3).

Table 3. Optimal rates of groundwater extraction from the First and Second Gravel Bed

Aquifer	Number of wells	Maximum withdrawal (m^3/day)	Safe yield (m^3/day)
	1	15.4	12
G-1	5	91.7	73
	9	151.8	121
	1	776	621
G-2	5	2059	1647
	9	2479	1983

CONCLUSION

Land subsidence due to groundwater overdraft is an environmental issue that has been affecting the Nobi Plain for several decades. Strict controls on water usage brought the problem under control; however, the spread of new economical activities especially towards urban outskirts indicate that the current regulations will soon be insufficient to meet the users' needs. The present work presented a numerical simulation that enhanced the understanding of the flow patterns, and predicted the safe yield of the First and Second aquifers for a sustainable development of the Aburashima region. In that case, the objective was to maximize the groundwater withdrawal while avoiding negative consequences. Results showed that water availability in the upper aquifer is somewhat restricted and therefore, abstraction from the lower unit would be more effective in terms of supply and relative cost.

Moreover, seasonal variations, and a higher demand for irrigation during summer have a significant impact on the exploitation of shallow wells, but are irrelevant for deep aquifers. When required, a feasible strategy to increase the abstraction rates is to distribute pumping wells in a network around the target site so as to cover larger areas of the aquifers.

It is anticipated that management authorities will utilize the present work in their efforts to adequate a sustainable supply to the present and future demand of groundwater. Although effective, restrictions on groundwater abstraction are only part of the solution since the continued growth of industrial and agricultural developments soon or later will derive in new conflicts between withdrawal and the ability to provide the resource. Control of groundwater depletion is not only a hydrogeologic problem, but it also involves political, socioeconomic, and technological factors. Therefore, a key point is to search for an integrated plan to manage the natural resources in the plain, which indefectibly will require the involvement of a number of governmental agencies and private parties. The cooperative work might be able to explore alternative measures to reduce the groundwater dependence, which indirectly would mitigate the subsidence threat. For example, it might be possible to evaluate the technical and economical feasibility of diverting surface waters for irrigation, construction of ponds and reservoirs for storage, increase in energy tariffs or taxes for larger groundwater consumers, restrict future industrial wells to deep aquifers, and implement subsides for cultivation of crops with low-water requirements.

ACKNOWLEDGEMENTS

The authors gratefully acknowledge Dr. Frank Columbus for the invitation to prepare the present paper.

REFERENCES

Don, NC., Araki, H., Yamanishi, H., and Koga, K (2005). Simulation of groundwater flow and environmental effects resulting from pumping. *Environ. Geol*. 47, 361-374.

Ebraheem, AM., Riad, S., Wycisk, P., and Sefelnasr, AM (2004). A local-scale groundwater flow model for groundwater resources management in Dakhla Oasis, SW Egypt. *Hydrogeology J*. 12 (6), 714-722.

Fetter, CW (2000). Applied Hydrogeology. 4th Edition. Prentice-Hall Publish, USA Geographical Survey of Japan (1997). *Digital Map 50 m Grid* (Elevation), Japan-II.

Groen, J., Velstra, A., and Meesters, AGCA (2000). Salinization processes in paleowaters in coastal sediments of Suriname: evidence from δ^{37} Cl analysis and diffusion modeling. *J. Hydrology* 234, 1-20.

Japan Meteorological Agency (2007). Climatic Report of the Japan Meteorological Agency, available from http://www.jma.go.jp/jma/indexe.html.

Larson, KJ., Başağaoğlu, H., and Mariño, MA (2001). Prediction of optimal safe ground water yield and land subsidence in the Los Baños-Kettleman City area, California, using a calibrated numerical simulation model. *J. Hydrology* 242, 79-102.

McDonald, MG and Harbaugh, AW (1988). A modular three-dimensional finite-difference ground-water flow model. US Geological Survey Techniques of Water Resources Investigations, Book 6.

Ministry of Economy, Trade and Industry of Japan (METI) (2006). Survey on the rational use of groundwater in the city of Kaizu, Gifu prefecture. *Report to the Bureau of the Ministry of Economy in Central Japan*, 117 pp (in Japanese).

Ministry of Economy, Trade and Industry of Japan (METI), (2007). Survey on the rational use of groundwater in the city of Kaizu, Gifu prefecture. *Report to the Bureau of the Ministry of Economy in Central Japan,* 73 pp (in Japanese).

Sakamoto, T., Takada, Y., Kuwahara, T. and Itoigawa, J (1986). Geology of the Nagoya-nambu district. Quadrangle series, scale 1:50,000, Kyoto (11) No. 32. *Geological Survey of Japan.* (in Japanese with English abstract).

Sakiyan, J. and Yazicigil, H (2004). Sustainable development and management of an aquifer system in western Turkey. *Hydrogeology J.* 12, 66-80.

Senthilkumar, M. and Elango, L (2004). Three-dimensional mathematical model to simulate groundwater flow in the lower Palar River basin, southern India. *Hydrogeology J.,* 12, 197-208.

Sophocleus, M (1998). Perspectives on sustainable development of water resources in Kansas, Bulletin 239, *Kansas Geological Survey*, Lawrence, Kansas, USA, pp 239.

Thornthwaite, CW. (1948). An approach toward a rational classification of climate. *Geol. Rev.* 35, 55-94.

Uchida, Y., Sakura, Y., and Taniguchi, M (2003). Shallow subsurface thermal regimes in major plains in Japan with reference to recent surface warming. *Phys. Chem. Earth 28*, 457-466.

Yamamoto, S. (1984). Case History No. 9.6. Nobi Plain, Japan. In J. F. Poland (Ed.), *Guidebook to studies of land subsidence due to ground-water withdrawal*, pp 195-204, UNESCO International Hydrological Programme, Working Group 8.4, USA.

Yamanaka, M. and Kumagai, Y (2006). Sulfur isotope constraint on the provenance of salinity in a confined aquifer system on the southwestern Nobi Plain, central Japan. *J. Hydrology* 325, 35-55.

Yoshida, F., Kurimoto, C., and Miyamura, M (1991). Geology of the Kuwana district. Quadrangle series, scale 1:50,000, Kyoto (11) No. 31. *Geological Survey of Japan.* (in Japanese with English abstract).

Waterloo Hydrogeologic (2005). *Visual MODFLOW* v. 4.1 User's Manual. Waterloo, Ontario.

In: Groundwater Research and Issues   ISBN: 978-1-60456-230-9
Editors: W. B. Porter, C. E. Bennington, pp. 45-104   © 2008 Nova Science Publishers, Inc.

Chapter 3

CHANGES OF COASTAL GROUNDWATER SYSTEMS IN RESPONSE TO LARGE-SCALE LAND RECLAMATION

Haipeng Guo and Jiu J. Jiao*

Department of Earth Sciences, The University of Hong Kong,
Hong Kong, P. R. China

ABSTRACT

Most large urban centers lie in coastal regions, which are home to about 25% of the world's population. The current coastal urban population of 200 million is projected to almost double in the next 20 to 30 years. This expanding human presence has dramatically changed the coastal natural environment. To meet the growing demand for more housing and other land uses, land has been reclaimed from the sea in coastal areas in many countries, including China, Britain, Korea, Japan, Malaysia, Saudi Arabia, Italy, the Netherlands, and the United States. Coastal areas are often the ultimate discharge zones of regional ground water flow systems. The direct impact of land reclamation on coastal engineering, environment and marine ecology is well recognized and widely studied. However, it has not been well recognized that reclamation may change the regional groundwater regime, including groundwater level, interface between seawater and fresh groundwater, and submarine groundwater discharge to the coast.

This paper will first review the state of the art of the recent studies on the impact of coastal land reclamation on ground water level and the seawater interface. Steady-state analytical solutions based on Dupuit and Ghyben-Herzberg assumptions have been derived to describe the modification of water level and movement of the interface between fresh groundwater and saltwater in coastal hillside or island situations. These solutions show that land reclamation increases water level in the original aquifer and pushes the saltwater interface to move towards the sea. In the island situation, the water divide moves towards the reclaimed side, and ground water discharge to the sea on both sides of the island increases. After reclamation, the water resource is increased because both recharge and the size of aquifer are increased.

E-mail address: jjiao@hku.hk. Tel.: (852) 2857 8246. Fax (852) 2517 6912. (Corresponding author)

This paper will then derive new analytical solutions to estimate groundwater travel time before and after reclamation. Hypothetical examples are used to examine the changes of groundwater travel time in response to land reclamation. After reclamation, groundwater flow in the original aquifer tends to be slower and the travel time of the groundwater from any position in the original aquifer to the sea becomes longer for the situation of coastal hillside. For the situation of an island, the water will flow faster on the unreclaimed side, but more slowly on the reclaimed side. The impact of reclamation on groundwater travel time on the reclaimed side is much more significant than that on the unreclaimed side. The degree of the modifications of the groundwater travel time mainly depends on the scale of land reclamation and the hydraulic conductivity of the fill materials.

1. INTRODUCTION

Most of the large urban centers are coastal and population in coastal areas is expanding rapidly. This expanding human presence has dramatically changed the coastal natural environment, which has made human as the most active geological agent in coastal areas. Coastal areas are faced with a challenge of lack of land due to increase in ecological, economic and social activities, which have led to an increase in pressure on coastal areas and in some cases led to the change of the coastal natural environment. Land reclamation, creating of new land in areas that were initially covered with water, has been a common practice to reduce the increasing pressure in coastal areas (Seasholes, 2003; Lumb, 1976; Suzuki, 2003; Stuyfzand, 1995). Coastal areas are usually the ultimate discharge zones of regional groundwater flow systems. Human activities such as large-scale land reclamation may have severe effect on coastal groundwater regimes.

In the early days, land was reclaimed by draining of swampy or seasonally submerged wetlands. In the coastal areas where intertidal zones have already been reclaimed, coastal reclamation usually means that shallow sea is landfilled extensively to create more useful land area. The analytical studies in this paper focus on the latter case. Land reclamation can increase the land for food production, create new investment into a city, and create new areas for urban development, easing housing shortages. From 1949 to 2001, about 22, 000 km^2 land, which is about half of the total tidal wetlands, along the coast of the Mainland China has been reclaimed. One third of urban area in Hong Kong is reclaimed from the sea. Land reclamation has been also very popular in many other countries including Japan, Korea, Malaysia, USA etc. For example, 45% of Korea's population lives in areas reclaimed from intertidal zone or shallow sea.

While reclamation satisfies the growing needs for land use, it also induces various engineering, environmental and ecological problems such as modification of coastal sediment deposition and erosion, coastal flooding, loss of the coastal habitat and change of marine environment. These problems have been widely discussed (Barnes, 1991; Noske, 1995; Ni et al., 2002; Terawaki et al., 2003). Coastal areas are often the ultimate discharge zones of regional ground water flow systems. Interaction between coastal groundwater and seawater always occurs (Fetter, 1972; Jiao and Tang, 1999), and is expected to be modified by land reclamation. Because the response of a groundwater regime to land reclamation may be slow and not so obvious, this phenomenon has been largely ignored. Stuyfzand (1995) discussed the impact of land reclamation on groundwater quality and future drinking water supply in the Netherlands. Mahamood and Twigg (1995) noted that the water table increased under areas related to reclaimed land in Bahrain Island of Saudi Arabia. A research team in the Department of Earth Sciences, the University of Hong Kong has carried out a research project aiming to understand the impact of large-scale land reclamation on physical and chemical process in coastal groundwater system. They conducted numerical modeling on the impact of land reclamation on groundwater flow and contaminant migration in Penny's Bay and Mid-Level areas, Hong Kong (Jiao, 2000; Jiao, 2002; Jiao et al., 2006). Jiao et al. (2001) presented some analytical solutions on the influence of reclamation on the water level changes in coastal unconfined aquifers. Jiao et al. (2005) hypothesized that there may be various physiochemical reactions between the fill materials and the original marine sediments in a reclamation site and these processes may result in chemicals which may have adverse effect to the coastal plants and organisms. Guo and Jiao (2007) presented some analytical solutions on impact of land reclamation on both the ground water level and the sea water interface with unconfined ground water conditions.

   Reclamation is expected to modify the groundwater travel time. Ground water travel time is defined as the time required for a unit volume of ground water to travel between two locations. The travel time is the length of the flow path divided by the velocity, where velocity is the average groundwater flux passing through the cross-sectional area of the geologic medium through which flow occurs, perpendicular to the flow direction, divided by the effective porosity along the flow path. If discrete segments of the flow path have different hydrologic properties the total travel time will be the sum of the travel times for each discrete segment (Code of Federal Regulations, 1988). Groundwater travel times have important implications for the management of water resources, evaluation of the groundwater availability and sustainability, and dating of groundwater. The assessment of groundwater travel times can be used to determine spreading of contaminated plume. The origin of groundwater particles can be tracked if their ages are known. In coastal aquifers, large amounts of dissolved and suspended materials are transported from the land to the sea together with terrestrial groundwater. Discharge of terrestrial groundwater derived from rainfall is a primary component of SGD (submarine groundwater discharge). Thus it is necessary to investigate sensitivity of the groundwater travel time in coastal aquifers.

   The ground water transit time is defined as the time interval of water flow from the recharge points to the discharge points (Etcheverry and Perrochet, 2000 ☐Chesnaux et al., 2005; Cornaton and Perrochet, 2005). Groundwater transit times have important applications and their sensitivity has been well investigated. Fórizs and Deák (1998) compared transit times calculated from stable oxygen isotope data and hydraulic modeling. They concluded a time series of groundwater sampling would be necessary when groundwater transit time was calculated from isotope data. Florea and Wicks (2001) constructed laboratory-scale karstic models to investigate transit times by controlling the length of the flow paths, the gradient, the recharge rate, the size of the conduits, and the geomorphic pattern of conduit layout. Their models provided reliable data that could be used in calibrating numerical models. A general theoretical framework to model complete groundwater age and transit time distributions at aquifer scale was presented by Cornaton and Perrochet (2005). The effect of aquifer structure and macro-dispersion on the distributions of age, life expectancy and transit time was discussed by means of analytical and numerical analysis of one- and two-dimensional theoretical flow configurations.

While numerical solutions for determining groundwater age and travel times are available (e.g., Varni and Carrera, 1998), they need more computational efforts and may be applicable for more complicated groundwater system. In many cases, analytical solutions can be used to assess travel times even though their application is limited by a certain number of assumptions. Gelhar and Wilson (1974) developed an analytical solution to calculate the groundwater transit time using the Dupuit assumption. Their analytical solution does not include the influence of the aquifer hydraulic conductivity. Amin and Campana (1996) presented a general lumped parameter mathematical model for estimating mean transit times in hydrologic systems characterized by different mixing regimes. Etcheverry and Perrochet (2000) calculated transit-time distributions based on the reservoir theory without considering the case of surface recharge. More recently, Chesnaux et al. (2005) developed a closed-form analytical solution for calculating groundwater transit times in unconfined aquifers, which included the influence of the hydraulic conductivity of the aquifer. Numerical verification was conducted to testify their analytical solution and comparison was made between their solution and that from Gelhar and Wison (1974). Chesnaux et al. (2005) concluded that the derived closed-form analytical solution was in excellent agreement with the numerical solution. Analytical solutions were also developed for computing the travel time for flow to a pumping well in unconfined aquifer without recharge (Simpson et al., 2003; Chapuis and Chesnaux, 2006).

An important class of ground water flow problems involves steady state, unconfined interface flow in coastal aquifers. Hydrologists have investigated the sea water interface for many decades. While analytical solutions do not directly solve "real-world" problems, they can serve as instructional tools, be used for first-cut engineering analysis and be used as benchmark problems for testing numerical algorithms (Bear et al., 1999). More than a century ago, a relationship resulting from Ghyben (1888) and Herzberg (1901) described that saltwater occurred underground at a depth below sea level of about forty times the freshwater head above the sea level. This relationship, known as Ghyben-Herzberg relation, has been widely used (Bear et al., 1999). With the sharp-interface assumption, many classical analytical solutions have been presented (Henry 1964; Glover 1964; Bear and Dagan 1964; Fetter 1972; Strack 1976). Glover (1964) used the method of complex variables to calculate the flow net and the position of the interface in an infinitely thick confined aquifer. Henry (1964) presented a solution for the position of the interface more precisely with the hodograph method. By comparing this solution with a model obtained under the Dupuit assumption (Fetter, 1994), Henry concluded that the use of the Dupuit assumption could produce sufficiently accurate results for most natural conditions. Fetter (1972) presented a general solution to determine the position of the saline interface under an oceanic island of any shape. A single-potential solution for regional interface problems in coastal aquifers was presented by Strack (1976), which was restricted to cases of steady flow with homogeneous isotropic permeability where the Dupuit assumption was valid.

Both Dupuit and Ghyben-Herzberg assumptions are also used in this paper. It is further assumed that the system is unconfined and all the ground water flow originates as precipitation recharge. The paper concerns only the long-term impact of large-scale land reclamation on coastal groundwater flow system and it is assumed that the ground water flow before reclamation is in a steady state and achieves another steady state after reclamation.

This paper consists of two parts. Based on the recent studies by Jiao et al (2001) and Guo and Jiao (2007), Part one presents the analytical solutions to study the long-term impact of land reclamation on both water level and saltwater interface in a coastal extensive land mass and an island with unconfined ground water conditions. Steady-state analytical solutions have been derived to describe the modification of water level and movement of the saltwater interface in coastal hillside or island situations. The derivation of the analytical solutions was presented in terms of water head. A far more advanced method in terms of the potential function (Strack, 1976) is also presented to obtain the analytical solutions (see appendix A and B). Variable-density flow and solute transport simulations, conducted by the numerical code FEFLOW, are used to evaluate the accuracy of the analytical solutions. A case study based on a major reclamation site in Shenzhen, China is presented to demonstrate the applicability of the analytical solutions.

Part two presents new analytical solutions for the travel time in coastal unconfined aquifers before and after reclamation for cases of both an extensive land mass and an island. Hypothetical examples are utilized to examine the difference of the new analytical solution and the solution developed by Chesnaux et al. (2005) which does not include the influence of the seawater-terrestrial groundwater interface. Sensitivity of travel times to the reclamation scale and hydraulic conductivity of the reclamation material is discussed. With the increase of the concern for pollution of aquifers, it becomes desirable to predict travel paths of contaminated water through the aquifer system. The streamline pattern will give a useful insight into spreading of contaminants. The impact of land reclamation on streamline pattern in the aquifer is investigated. Furthermore, the changes of the flow-velocity distribution in response to land reclamation are discussed.

2. Impact of Land Reclamation in an Extensive Land Mass

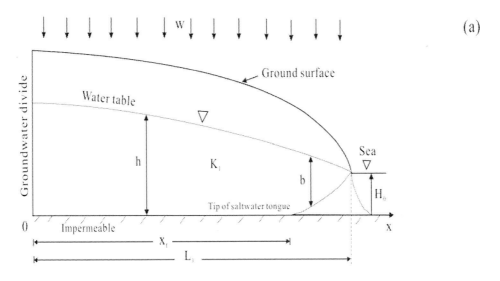

Figure 1. (Continued on next page.)

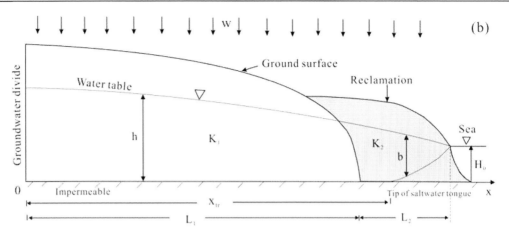

Figure 1. Sketch of an unconfined aquifer system and the seawater-terrestrial groundwater interface in a coastal extensive land mass (a) before reclamation and (b) after reclamation (Guo and Jiao, 2007).

Impact of land reclamation on groundwater flow in an extensive land mass was investigated by Jiao et al. (2001) and Guo and Jiao (2007). Problems of combined shallow unconfined interface flow and unconfined flow often occur in coastal aquifers. Figure 1a shows a coastal unconfined ground water system receiving uniform vertical recharge w. The fresh groundwater and the sea water near the coast are separated by a saltwater interface. The hydraulic conductivity of the unconfined aquifer is K_1. The distances from the water divide to the coastline and the tip of the saltwater tongue are represented as L_1 and x_t, respectively (Figure 1a). The head in the fresh water is h above the horizontal impermeable bottom of the aquifer, which is H_0 below sea level. After reclamation (Figure 1b), the coastline moves toward the sea by a distance of L_2, which is hereafter called reclamation length, and the hydraulic conductivity of the fill materials is K_2 (the boundary between K_1 and K_2 is approximated as vertical). The distance from the water divide to the post-reclamation tip is assumed to be x_{tr}. This study focuses primarily on changes of coastal groundwater systems in response to large-scale land reclamation. In this case, it is assumed that the post-reclamation tip of the salt water tongue is located in the reclaimed land when steady-state flow conditions are achieved. The conceptual models in Figure 1 are similar to those presented by Guo and Jiao (2007).

2.1. Impact of Reclamation on Ground Water Level and Saltwater Interface

2.1.1. Solution for Ground Water Level and Saltwater Interface
According to Ghyben-Herzberg assumption (Bear et al. 1999), the seawater and fresh groundwater are assumed to be separated by a sharp interface rather than by a transition zone. The saltwater interface can be regarded as a fixed impermeable boundary at a distance below sea level equal to

$$\bar{z} = \frac{\rho_f (h - H_0)}{\rho_s - \rho_f} \tag{1}$$

where ρ_f and ρ_s are the densities of ground water and seawater. As showed in Figure 1a, the depth of flow in the Dupuit equation, b, is

$$b = h \qquad\qquad (0 \le x \le x_t) \qquad\qquad (2)$$

$$b = \overline{z} + h - H_0 = \frac{\rho_s(h - H_0)}{\rho_s - \rho_f} \qquad\qquad (x_t \le x \le L_1) \qquad\qquad (3)$$

The depth of the saltwater interface below sea level $\overline{z}$ is equal to H_0 at $x = x_t$, which leads to

$$h = \frac{\rho_s}{\rho_f} H_0 \qquad\qquad (x = x_t) \qquad\qquad (4)$$

The Dupuit assumption, which states that equipotential surfaces are essentially vertical and the flow horizontal, is a most powerful tool for treating unconfined or unconfined interface flow. Ground water flow in the unconfined aquifer is assumed to satisfy Dupuit assumption and can be described as:

$$\frac{d}{dx}\left(Kb\frac{dh}{dx}\right) + w = 0 \qquad\qquad (5)$$

where b is as defined by expressions (2) and (3), and K is the hydraulic conductivity of the aquifer. Guo and Jiao (2007) presented analytical solutions for the water level and the positions of the saltwater interface before and after reclamation. The main solutions are as follows:

(I)   before reclamation

$$h = \sqrt{\frac{w}{K_1}(L_1^2 - x^2) + \frac{\rho_s}{\rho_f}H_0^2} \qquad\qquad (0 \le x \le x_t) \qquad\qquad (6)$$

$$h = \sqrt{\frac{w(\rho_s - \rho_f)}{K_1\rho_s}(L_1^2 - x^2) + H_0} \qquad\qquad (x_t \le x \le L_1) \qquad\qquad (7)$$

$$x_t = \sqrt{L_1^2 - \frac{K_1(\rho_s^2 - \rho_s\rho_f)}{w\rho_f^2}H_0^2} \qquad\qquad (8)$$

(II)   after reclamation

$$h = \sqrt{w\left[\frac{1}{K_1}(L_1^2 - x^2) + \frac{1}{K_2}(2L_1 + L_2)L_2\right] + \frac{\rho_s}{\rho_f}H_0^2} \qquad (0 \le x \le L_1)$$

(9)

$$h = \sqrt{\frac{w}{K_2}\left[(L_1 + L_2)^2 - x^2\right] + \frac{\rho_s}{\rho_f}H_0^2} \qquad (L_1 \le x \le x_{tr})$$

(10)

$$h = \sqrt{\frac{w(\rho_s - \rho_f)}{K_2\rho_s}\left[(L_1 + L_2)^2 - x^2\right]} + H_0 \qquad (x_{tr} \le x \le L_1 + L_2)$$

(11)

$$x_{tr} = \sqrt{(L_1 + L_2)^2 - \frac{K_2(\rho_s^2 - \rho_s\rho_f)}{w\rho_f^2}H_0^2}$$

(12)

2.1.2. Numerical Verification and Comparison

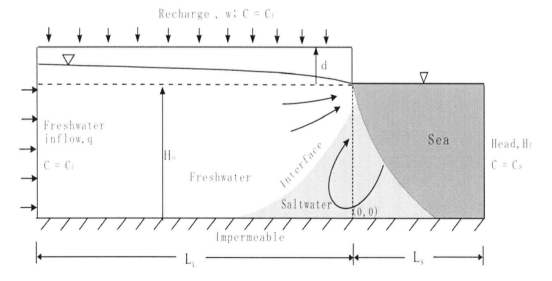

Figure 2. Schematic diagram of the geometry and boundary conditions used in the numerical verification.

In reality, the seawater-terrestrial groundwater interface is a transition zone so that numerical experiments with a variably-density flow model need to be conducted to check the validity of the derived analytical solutions. The analytical solutions for the water table and the seawater-freshwater interface are here tested numerically for a field-scale problem using the FEFLOW code (Diersch, 2002).

Geometry, Aquifer Properties, Boundary Conditions, and Numerical Strategy

Figure 2 shows the geometry and boundary conditions that are considered to verify the previous analytical solutions. It represents a vertical section through a typical coastal unconfined aquifer. The notations L_L and L_S are the landward and seaward lengths. L_L and L_S were determined experimentally so that they didn't influence the shape of the mixing zone significantly (Smith, 2004). The notation d is the distance from the sea level to the land surface. The sea level is assumed to be H_0 above the impermeable bottom boundary. The coordinate origin is set to be on the bottom boundary with the same x coordinate of the shoreline (see Figure 2).

The base of the aquifer at $z = 0$ is assumed to be a no-flow boundary. Neuman boundary conditions are specified along the left model boundary and the landward portion of the top boundary, which respectively represent the terrestrial groundwater inflow and the local recharge. The boundary conditions can be described as follows:

$$C(-L_L, z) = C_f \qquad\qquad 0 \le z \le H_0 \tag{13}$$

$$Q(-L_L, z) = w(L_1 + L_2 - L_L) \quad 0 \le z \le H_0 \tag{14}$$

$$C(x, H_0 + d) = C_f \qquad\qquad -L_L \le x \le 0 \tag{15}$$

$$N(x, H_0 + d) = w \qquad\qquad -L_L \le x \le 0 \tag{16}$$

where C_f is the salt concentration of the fresh groundwater, Q is the discharge per unit width (L^2/T), N is the uniform recharge across land portion of the upper boundary (L/T), and w is the infiltration rate used in the numerical solution (L/T). L_1 and L_2 are the distance from the water divide to the original coastline and the reclamation length, respectively (Figure 1).

The seabed is assumed to be horizontal in all simulations and the sea level static so that influences of the tidal fluctuations, waves, and the seasonal water exchange between the freshwater and the seawater are neglected. The simplification of the surface water body to be non-penetrating has been adopted in many works (Townley and Davidson, 1988; Nield et al.,1994; Smith and Turner, 2001; Smith, 2004). The boundary conditions along the seabed portion of the top boundary and the right boundary can be described as follows:

$$H_f(x, H_0) = 0 \qquad\qquad 0 \le x \le L_S \tag{17}$$

$$C(x, H_0) = C_S \qquad\qquad 0 \le x \le L_S \tag{18}$$

$$H_f(L_S, z) = z + \frac{\rho_s}{\rho_f}(H_0 - z) \qquad 0 \le z \le H_0 \tag{19}$$

$$C(L_S, z) = C_S \qquad\qquad 0 \le z \le H_0 \qquad\qquad (20)$$

where H_f is the hydraulic head (L) expressed at the fresh groundwater density ρ_f (M/L^3) ρ_s is the density of sea water, C is the salt concentration (M/L^3), and C_S is the salt concentration of the seawater. The so-called "boundary constraint" condition available in FEFLOW simulator is utilized to achieve the described mass boundary.

Three hypothetical cases were simulated to verify the accuracy of the analytical solutions (Table 1). An iterative solution for steady flow and mass transport was obtained for case 1. However, the solutions could not converge for cases 2 and 3, then transient model runs were performed to achieve quasi-steady states.

The simulation parameters are presented in Table 2. These parameters are based on those used by Smith (2004) for similar coastal groundwater problems.

The numerical solution strategy adopted in the simulation is listed in Table 3. The Boussinesq approximation is applied so that density variations are included by the buoyancy term of the Darcy equation only. This approximation can also simplify the mathematical model of the variable density flow. In addition, the so-called no-upwinding approach is used to keep numerical dispersion as low as possible (Kolditz et al., 1998). Furthermore, the influence of the viscosity on the hydraulic conductivity is neglected in this study.

Table 1. FEFLOW Model Inputs for the Hypothetical Simulation Cases

Case	K_1 m/day	K_2 m/day	L_L m	L_S m	w m/day	L_1 m	L_2 m	H_0 m	Description
Case 1	3	---	200	50	0.0005	1000	---	20	Before Reclamation
Case 2	3	5	200	50	0.0005	1000	500	20	After Reclamation
Case 3	3	10	400	100	0.0005	1000	500	20	After Reclamation

Table 2. Simulation Parameters for the Numerical Verification

Symbol	Quantity	Value	Unit
D_m	coefficient of molecular diffusion	8.64×10^{-5}	m^2/day
φ	Porosity	0.3	---
C_f	salt concentration of fresh groundwater	0	mg/L
C_S	salt concentration of seawater	35,000	mg/L
$\overline{\alpha}$	density difference ratio	0.025	---
α_L	longitudinal dispersivity of solute	5	m
α_T	transverse dispersivity of solute	0.5	m

Table 3. Numerical Specifications in FEFLOW

Topic	*Quantity*
Mesh type	3-noded triangle elements (82,162 to 555,696)
Finite element scheme	no upwinding (best-accurate Galerkin-based formulation)
Problem class	combined flow and mass transport
Time stepping scheme	automatic time step control via predictor-corrector schemes (AB/TR time integration scheme)
Mass transport equation	convective form transport
Iterative solver	preconditioned conjugate gradient PCG
Density coupling	Boussinesq approximation
Evaluation of element integral	standard Gauss quadrature
Velocity approximation	improved consistent velocity approximation (by Frolkovic-Knabner)

Comparison of the results from numerical simulation and those from the analytical solution

Three hypothetical cases (Table 1) are utilized to verify the accuracy of the analytical solutions. Figure 3 depicts white streamlines in the mixing zone superimposed on shaded concentration fringes, which represent the density-driven seawater circulation in the coastal aquifer. This kind of circulation is induced by the displacement of the freshwater by the seawater and the salt dispersion in the mixing zone. A discharge zone forms near the coastline (Figure 3), inducing a natural, realistic shape of the transition zone. It can be inferred from Equation (12) that the post-reclamation seawater intrusion length, $L_1+L_2-x_{tr}$, increases with the hydraulic conductivity of the reclamation material, K_2. The transition zone in Figure 3(c) tends to be more landward than that in Figure 3(b) because the hydraulic conductivity of the aquifer is increased from 5m/day to 10 m/day.

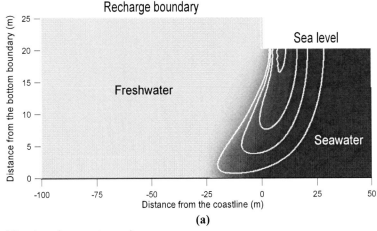

Figure 3. (Continued on next page.)

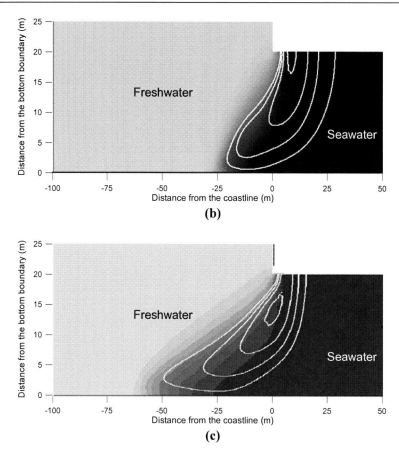

(b)

(c)

Figure 3. Steady-state simulation results for the hypothetical examples. White streamlines superimposed on shaded concentration fringes show the seawater circulations. Only results of a horizontal range of 150m besides the coastline are presented in order to compare the simulated mixing zones in different cases. (a) Hydraulic conductivity of the aquifer is 3m/day and no reclamation occurs. (b) Hydraulic conductivities of the original aquifer and the reclaimed aquifer are 3m/day and 5m/day, respectively. (c) Hydraulic conductivities of the original aquifer and the reclaimed aquifer are 3m/day and 10m/day, respectively.

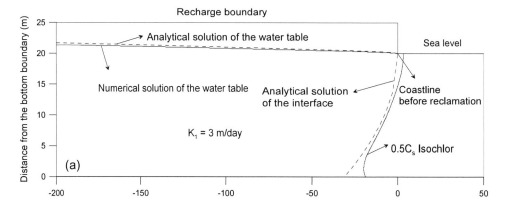

Figure 4. (Continued on next page.)

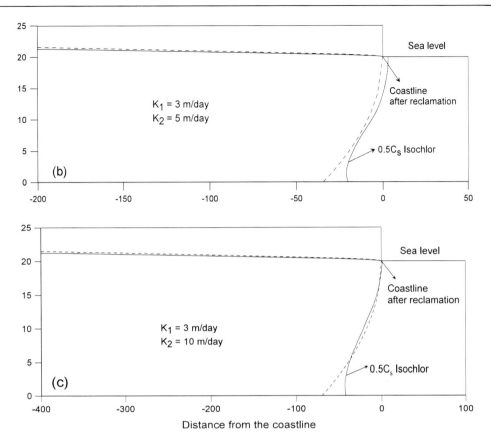

Figure 4. Steady state 0.5C$_S$ isochlors compared with saltwater interfaces obtained from the analytical solutions for three different cases. (a) Hydraulic conductivity of the aquifer is 3m/day and no reclamation occurs. (b) Hydraulic conductivities of the original aquifer and the reclaimed aquifer are 3m/day and 5m/day. (c) Hydraulic conductivities of the original aquifer and the reclaimed aquifer are 3m/day and 10m/day.

Figure 4 compares the 0.5Cs isochlors at steady state for the three cases in Table 1 with the corresponding saltwater interfaces obtained from the analytical solutions. The accuracy of the analytical solution is somewhat limited by assumptions such as Dupuit-type flow and the Ghyben-Herzberg relation. Because of these assumptions, the saltwater interfaces from the analytical solutions do not completely agree with the 0.5Cs isochlors produced by the density-driven transport model. The 0.5 Cs isochlors are a little further seaward than the sharp interface obtained from the analytical solutions. Comparisons of the water tables between the numerical solutions and the analytical solutions show that the water tables calculated from the analytical solutions are slightly higher than those from the numerical solutions. However, these differences are not very significant and the maximal difference is 0.5m, which is only 2.5% of the elevation of the sea level in relation to the bottom boundary. Thus the simplifying assumptions used in the analytical studies are testified to be reasonable.

2.1.3. Discussion of the Analytical Solutions

Change of the water level in the original aquifer

The change of the water table in the original aquifer between $x = 0$ and $x = L_1$ induced by reclamation can be calculated from Equations (6), (7) and (9), which yields

$$\Delta h = \sqrt{w\left[\frac{1}{K_1}(L_1^2 - x^2) + \frac{1}{K_2}(2L_1 + L_2)L_2\right] + \frac{\rho_s}{\rho_f}H_0^2} - \sqrt{\frac{w}{K_1}(L_1^2 - x^2) + \frac{\rho_s}{\rho_f}H_0^2}$$

$$(0 \leq x \leq x_t) \tag{21}$$

$$\Delta h = \sqrt{w\left[\frac{1}{K_1}(L_1^2 - x^2) + \frac{1}{K_2}(2L_1 + L_2)L_2\right] + \frac{\rho_s}{\rho_f}H_0^2} - \sqrt{\frac{w(\rho_s - \rho_f)}{K_1\rho_s}(L_1^2 - x^2) - H_0}$$

$$(x_t < x \leq L_1) \tag{22}$$

Equations (21) and (22) can be utilized to analyze sensitivity of the ground water table change to the hydraulic conductivity of the reclamation material and the scale of reclamation. The change of the water level at the original coastline, which is the maximum change in the domain, can be readily obtained by setting $x = L_1$ in Equation (22), which leads to

$$\Delta h = \sqrt{\frac{w}{K_2}(2L_1 + L_2)L_2 + \frac{\rho_s}{\rho_f}H_0^2} - H_0 \tag{23}$$

Equation (23) indicates the maximum buildup of the water level at the original coastline is independent of K_1, and increases with L_1, L_2, and the ratio between the recharge rate w and K_2. For a specific coastal area, the parameters L_1 and w are fixed so that the water level buildup at the original coastline mainly depends on the hydraulic conductivity of the reclamation material K_2 and the reclamation length L_2. The reclamation tends to have a significant damming effect when L_1 is great and K_2 is low. This conclusion is the same as that of the study by Jiao et al (2001) which ignored the seawater interface.

Displacement of the tip of the saltwater tongue

Due to reclamation the position of the tip of the saltwater tongue moves towards the sea by a distance of $x_{tr} - x_t$, which can be obtained from Equations (8) and (12):

$$x_{tr} - x_t = \sqrt{(L_1 + L_2)^2 - \frac{k_2(\rho_s^2 - \rho_s\rho_f)}{w\rho_f^2}H_0^2} - \sqrt{L_1^2 - \frac{k_1(\rho_s^2 - \rho_s\rho_f)}{w\rho_f^2}H_0^2}$$

$$\tag{24}$$

For a particular coastal area, parameters L_1, K_1, w, and H_0 are fixed, the change of the tip of the saltwater tongue depends mainly on L_2 and K_2. As mentioned early, this study assumes that the tip of the saltwater tongue after reclamation is located in the reclaimed land when steady state flow conditions are reached. In this case, the value $x_{tr} - x_t$ is always positive, i.e. the tip of the saltwater interface will be pushed seaward after reclamation. Equation (24) shows that displacement of the tip, $x_{tr} - x_t$, is significant if the reclamation length L_2 is great and the hydraulic conductivity of the reclamation material K_2 is low.

2.1.4. Hypothetical Example

The impact of the reclamation on the ground water level and the seawater-freshwater interface is discussed here with a hypothetical example. Assume that the saturated hydraulic conductivity K_1 of the aquifer is 0.1 m/day and the distance from the ground water divide to the original coastline, L_1, is 1000m. The densities of seawater and freshwater are 1.025 g/cm^3 and 1.000 g/cm^3, respectively. The infiltration rate is 0.0005 m/day and H_0 =20m.

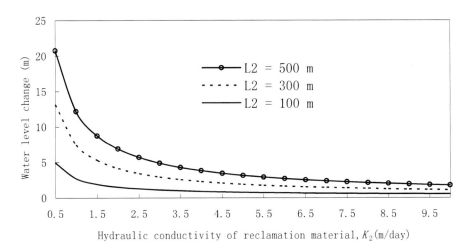

Figure 5. Change of ground water level at the original coastline with hydraulic conductivity of the reclamation materials when K_1=0.1m/day, L_1=1000m, ρ_s =1.025g/cm^3, ρ_f =1.000g/cm^3, w=0.0005m/day, and H_0=20m.

The land reclamation will increase the water level in the original aquifer and push the saltwater interface seaward. A comparison of predicted water levels at the original coastline as a function of hydraulic conductivity and reclamation length shows that the water level rise at the original coastline after reclamation decreases with K_2 and increases with L_2 (Figure 5). The buildup of the water level at the original coastline can be great and is sensitive to K_2 when K_2 is low. When K_2 is greater than 5 m/day, the water-level rise tends to be small and insensitive to K_2.

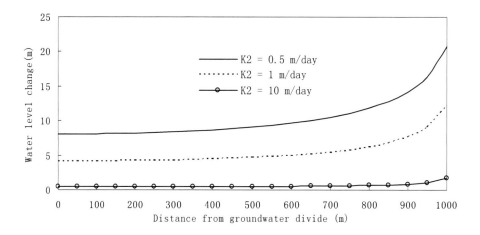

Figure 6. Change of ground water level in the original unconfined aquifer with distance from the ground water divide when K_1=0.1m/day, L_1=1000m, L_2=500m, ρ_s =1.025g/cm³, ρ_f =1.000g/cm³, w=0.0005m/day, and H_0=20m.

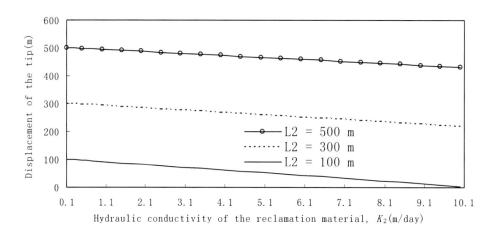

Figure 7. Displacement of the tip of the saltwater tongue with hydraulic conductivity of the reclamation materials for different reclamation lengths when K_1=0.1m/day, L_1=1000m, ρ_s =1.025g/cm³, ρ_f =1.000g/cm³, w=0.0005m/day, and H_0=20m (Guo and Jiao, 2007).

Figure 6 shows the change of ground water level in the original unconfined aquifer with distance from the ground water divide when L_2=500m. The rise of the water level increases with the distance from the ground water divide and decreases with hydraulic conductivity of the fill materials. For any fixed value of K_2, the water level change is sensitive to distance from the water divide and can be significant at positions close to the original coastline. The water-level rise is much more sensitive to K_2, and the maximum rise is located at the original coastline. With L_2 = 500 m the water level rise at the original coastline is 20.7 m for K_2 = 0.5 m/day and only 1.7 m for K_2 = 10 m/day.

Figure 7 shows the displacement of the tip of the saltwater tongue, which is defined as $x_{tr}-x_t$, as a function of L_2 and K_2. The displacement of the tip decreases with K_2 and increases with L_2, and is usually less than L_2 when $K_2 > K_1$. The displacement is slightly greater than L_2 when $K_2 = K_1$ because the tip is pushed much more seaward due to the increased recharge in the reclaimed land. The displacement of the tip decreases almost linearly with K_2, indicating that the post-reclamation saltwater tongue can be significantly longer than the initial one when K_2 is great. After reclamation, the saltwater interface is pushed seaward, which may benefit groundwater wells screened close to the saltwater interface.

2.1.5. Application: a Case Study in Shenzhen, China

The study site covers a major reclamation area to the southwest of Shenzhen, China (see Figure 8). The rapid urban development has led to a sharp increase in the demand for usable land in this area during the past 20 years, and large-scale land reclamation by filling the shallow sea has been a common practice to ease this demand. Figure 8 shows the original coastline in 1983 and the coastline in 2000 around the Deep Bay, Shenzhen. The water divide is located about 4.2 km north of the reclamation area and lies along topographical peaks and ridges. Between 1983 and 2000, the coastline has been pushed towards the sea by a distance of approximately 700 m due to a series of reclamation projects.

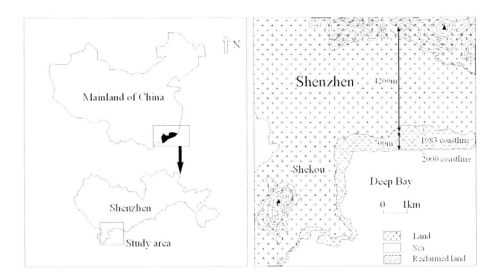

Figure 8. Change of the coastline due to reclamation in Shenzhen, China between year 1983 and year 2000.

The study area around Deep Bay is underlain mainly by weathered igneous rock. Ding (2006) analyzed the measured hydraulic conductivity values of the weathered granite and concluded that average hydraulic conductivity value was of the order of 10^{-6} m/s. In the analytical study, the hydraulic conductivity of the original aquifer K_1 is assumed to be 1.0×10^{-6} m/s. The approximate hydraulic conductivity of the fill material in this area is of the order of 10^{-4} m/s. The climate of this area is subtropical humid with hot wet summer and mild dry winter, and the average annual rainfall is about 1837 mm. The recharge coefficient is here taken as 0.1 over a long-term basis, which leads to an annual infiltration rate of about

184mm. The average reclamation thickness in Deep Bay is about 10 m. According to Equations (22) and (24), the estimated values are very sensitive to hydraulic conductivity of the fill material K_2, which is uncertain in the field. If $K_2 = 2.0 \times 10^{-4}$ m/s, the estimated change of the water level at the original coastline is 6.9m. Because of the reclamation, the saltwater tongue will be pushed to the new coastline, and the displacement of the tip of the tongue is estimated to be 691m based on (24). The change of the water level at the original coastline and the displacement of the tip are estimated to be 3.3m and 677m when K_2 is increased to be 5.0×10^{-4} m/s.

2.2. Impact of Reclamation on Ground Water Travel Time

According to the Dupuit theory, the flow in the unconfined aquifer is considered horizontal and one dimensional, and the ground water discharge per unit width of aquifer is defined as:

$$Q_x = -Kb \frac{dh}{dx} = q_x b \qquad (25)$$

where Q_x is the ground water discharge per unit width (L^2/T), K is the saturated hydraulic conductivity(L/T), b is the thickness of the fresh groundwater (L) , h is the hydraulic head (L), and q_x is specific discharge or Darcy flux (L/T).

In the unconfined systems (Figure 1), infiltration occurs at a constant rate w along the entire upper boundary of the aquifer. The ground water discharge per unit width of aquifer affected by a uniform recharge is

$$Q_x = w\, x \qquad (26)$$

Since the vertical component of the flow velocity is neglected, the ground water velocity v (L/T) can be expressed as:

$$v_x = \frac{dx}{dt} = \frac{q_x}{n_e} \qquad (27)$$

where n_e is the effective porosity of the porous medium. Closed-form solutions for travel time t (T) of a water particle to move between two arbitrary positions x_i (at time t_i) and x (at time t) in the unconfined aquifer can be obtained by combining Equations (25), (26) and (27), which leads to

$$\int_{t_i}^{t} dt = \frac{n_e}{w} \int_{x_i}^{x} \frac{b(\tau)}{\tau} d\tau \qquad (28)$$

Based on Equation (28), Chesnaux et al. (2005) developed an analytical solution to calculate ground water transit time in a Dupuit-type flow system bounded by lakes or rivers (Figure 9). Figure 9 shows the unconfined system used for developing the analytical solution. The left-hand boundary is groundwater divide, and ground water discharges through the right-hand fixed-head boundary. The solution of the transit time from an arbitrary point at which x = x_i (assuming the travel path begins at the water table) can be expressed as:

$$t(x_i) = n_e \sqrt{\frac{\zeta}{wK}} (L' \sqrt{\frac{1}{L'^2} - \frac{1}{\zeta}} - x_i \sqrt{\frac{1}{x_i^2} - \frac{1}{\zeta}} + \ln \frac{\frac{\sqrt{\zeta}}{x_i} + \sqrt{\frac{\zeta}{x_i^2} - 1}}{\frac{\sqrt{\zeta}}{L'} + \sqrt{\frac{\zeta}{L'^2} - 1}})$$

(29)

where $\zeta = L'^2 + Kh_L^2 / w$, L' represents the length of the unconfined aquifer system, and $h_{L'}$ is the fixed-head of the right-hand boundary.

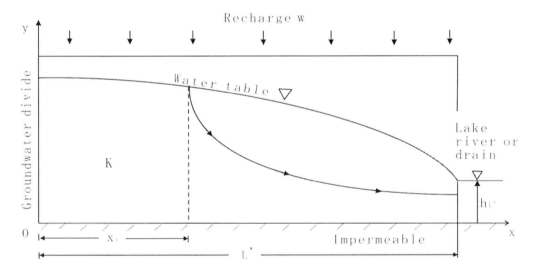

Figure 9. The unconfined system used for developing the analytical solution by Chesnaux et al. (2005). The left-hand boundary is groundwater divide, and ground water discharges through the right-hand fixed-head boundary.

2.2.1. Analytical Solution for the Ground Water Travel Time before Reclamation

Analytical solution

When seawater-freshwater interface is considered, the thickness of the cross-sectional area through which flow occurs will not be always equal to the water head. Change of thickness of freshwater with distance towards the left boundary can be obtained from Equations (2), (3), (6) and (7). The final solutions are

$$b = h = \sqrt{\frac{w}{K_1}(L_1^2 - x^2) + \frac{\rho_s}{\rho_f} H_0^2} \qquad (0 \le x \le x_t) \qquad (30)$$

$$b = \frac{\rho_s(h - H_0)}{\rho_s - \rho_f} = \sqrt{\frac{w\rho_s}{K_1(\rho_s - \rho_f)}(L_1^2 - x^2)} \qquad (x_t \le x \le L_1) \qquad (31)$$

Substituting Equations (30) and (31) into Equation (28), one has:

$$\int_{t_i}^{t} dt = \frac{n_e}{\sqrt{wK_1}} \int_{x_i}^{x} \frac{\sqrt{\omega^2 - x^2}}{x} dx \qquad (0 \le x \le x_t) \quad (32)$$

$$\int_{t_i}^{t} dt = n_e \sqrt{\frac{\rho_s}{wK_1(\rho_s - \rho_f)}} \int_{x_i}^{x} \frac{\sqrt{L_1^2 - x^2}}{x} dx \quad (x_t \le x \le L_1)$$
$$(33)$$

where $\omega = \sqrt{L_1^2 + \frac{\rho_s K_1}{\rho_f w} H_0^2}$

The solution to Equations (32) and (33) leads to:

$$\int_{t_i}^{t} dt = n_e \sqrt{\frac{1}{wK_1}} \left[\sqrt{\omega^2 - x^2} - \sqrt{\omega^2 - x_i^2} - \omega \ln\left(\frac{\omega + \sqrt{\omega^2 - x^2}}{\omega + \sqrt{\omega^2 - x_i^2}} \frac{x_i}{x}\right) \right]$$
$$(0 \le x \le x_t) \qquad (34)$$

$$\int_{t_i}^{t} dt = n_e \sqrt{\frac{\rho_s}{(\rho_s - \rho_f)wK_1}} \left[\sqrt{L_1^2 - x^2} - \sqrt{L_1^2 - x_i^2} - L_1 \ln\left(\frac{L_1 + \sqrt{L_1^2 - x^2}}{L_1 + \sqrt{L_1^2 - x_i^2}} \frac{x_i}{x}\right) \right]$$
$$(x_t \le x \le L_1) \qquad (35)$$

Integrating Equations (34) and (35) with respect to x from x_0 to x_1 yields the travel time of water flow from one position ($x = x_0$) to another position ($x = x_1$) in the unconfined aquifer before reclamation. If $x_1 = L_1$, the resultant travel time is the transit time of a particle of water originating at $x = x_0$ before reclamation. The total travel time will be the sum of the travel times for each discrete segment if discrete segments of the flow path have different aquifer properties.

Comparison of the solutions with and without considering seawater interface
A hypothetical example is used here to compare the new analytical solution with the transit time solution developed by Chesnaux et al. (2005), which didn't include the influence of seawater-freshwater interface and can be expressed as Equation (29). Assume the distance from the ground water divide to the coastline is 1000m, and the effective porosity of the aquifer is 0.3. The difference between the groundwater travel time with and without the seawater-freshwater interface is defined as

$$\Delta t = t_{New} - t_{Chesnaux} \tag{36}$$

where t_{New} and $t_{Chesnaux}$ represent the groundwater travel time calculated from the new analytical solution in this paper and from Equation (29) developed by Chesnaux et al. (2005), respectively.

Figure 10 shows the differences (defined by Equation 36) of transit times to the sea from different positions in the aquifer with distance from the water divide for different sea levels when w = 0.0005 m/day and K_1 = 3 m/day. The calculated water level becomes higher when the impact of the seawater-freshwater interface is considered. Thus, the thickness of the freshwater between the water divide and the tip of the saltwater tongue will be greater, thereby decreasing the flow velocity. However, the groundwater flow velocity tends to be greater in the area between the tip and the coastline due to decrease of the freshwater thickness in this area. Therefore, as shown in Figure 10, differences of transit times decrease with the distance from groundwater divide, and tend to be positive close to the water divide and negative near the coastline. When H_0 is greater, differences of the transit times will be more significant because the seawater intrusion length tends to be greater according to Equation (8). With H_0 = 30 m the difference is 188.1 days for x = 50m and -424.8 days for x=950m.

Figure 11 shows differences of transit times to the sea from different positions in the aquifer with distance from the water divide for different infiltration rates when K_1 = 3 m/day and H_0 =20m. The curves in Figure 11 show the similar trends as those in Figure 10. According to Equation (8), the influence of the seawater intrusion will be great when the recharge rate w is low. Therefore, differences of the transit times to the sea tend to be more significant when the recharge rate w is low. With w = 0.0005 m/day the difference is 260.1 days for x = 50m and -118.5 days for x=950m.

Figure 12 shows differences of the transit times with distance from the water divide for different hydraulic conductivities of the aquifer when w = 0.0005 m/day and H_0 =20m. According to Equation (8), the seawater intrusion length, L_1-x_1, becomes longer when K_1 is increased. In this case, the flow-velocity increase due to decrease of the freshwater thickness between the tip and the coastline will be more significant and be in a greater area. Thus, differences of transit times decrease with K_1. With the distance from the groundwater divide increasing, the impact of the flow-velocity increase close to the coastline on the transit time will become stronger. At the same time, the influence of the flow-velocity decrease due to rise of the water table away from the coastline will diminish. Therefore, differences of the transit times decrease with the distance from the groundwater divide.

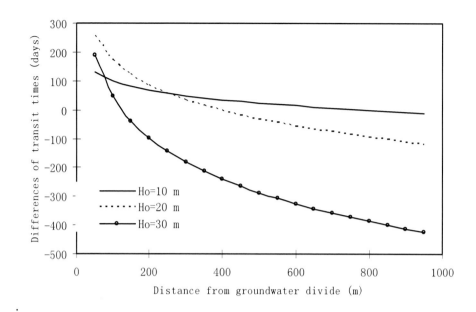

Figure 10. Differences of transit times to the sea from different positions in the aquifer with distance from the water divide for different sea levels when K_1=3m/day, L_1=1000m, ρ_s =1.025g/cm^3, ρ_f =1.000g/cm^3, and w=0.0005m/day.

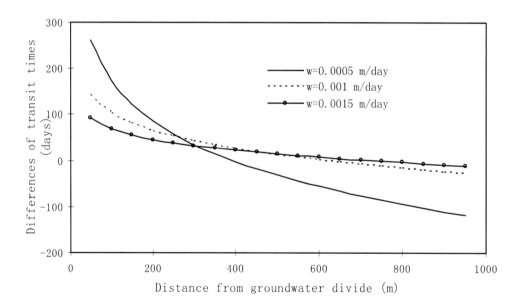

Figure 11. Differences of the transit times to the sea from different positions in the aquifer with distance from the left water divide for different infiltration rates when K_1=3m/day, L_1=1000m, ρ_s =1.025g/cm^3, ρ_f =1.000g/cm^3, and H_0 =20m.

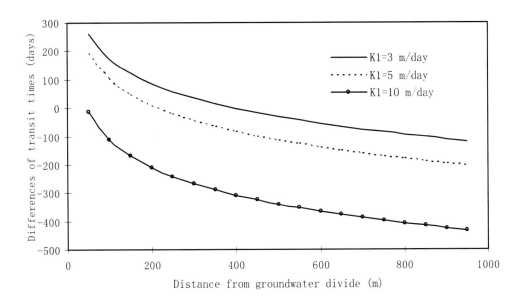

Figure 12. Differences of the transit times to the sea from different positions in the aquifer with distance from the water divide for different hydraulic conductivities of the aquifer when L_1=1000m, ρ_s =1.025g/cm^3, ρ_f =1.000g/cm^3, w = 0.0005 m/day, and H_0 =20m.

2.2.2. Analytical Solution for the Ground Water Travel Time after Reclamation

Analytical solution

After reclamation the coastline is pushed seaward by a distance of L_2 so that an additional aquifer forms and rain recharge takes place over a larger area (Figure 1b). The change of the freshwater thickness with distance from the left boundary can be obtained from Equations (9), (10) and (11) as follows:

$$b = h = \sqrt{w\left[\frac{1}{K_1}(L_1^2 - x^2) + \frac{1}{K_2}(2L_1 + L_2)L_2\right] + \frac{\rho_s}{\rho_f}H_0^2} \qquad (0 \le x \le L_1)$$

$$(37)$$

$$b = h = \sqrt{\frac{w}{K_2}\left[(L_1 + L_2)^2 - x^2\right] + \frac{\rho_s}{\rho_f}H_0^2} \qquad (L_1 \le x \le x_{tr})$$

$$(38)$$

$$b = \frac{\rho_s(h - H_0)}{\rho_s - \rho_f} = \sqrt{\frac{w\rho_s}{K_2(\rho_s - \rho_f)}\left[(L_1 + L_2)^2 - x^2\right]} \qquad (x_{tr} \le x \le L_1 + L_2)$$

$$(39)$$

Equations (37), (38) and (39) can be used to describe relation between thickness of the vertical cross section through which water flow occurs and the distance towards the water divide. Substituting these equations to equation (28) respectively and integrating the derived equations, the following equations are obtained:

$$\int_{t_i}^{t} dt = n_e \sqrt{\frac{1}{wK_1}} \left[\sqrt{\varphi^2 - x^2} - \sqrt{\varphi^2 - x_i^2} - \varphi \ln \left(\frac{\varphi + \sqrt{\varphi^2 - x^2}}{\varphi + \sqrt{\varphi^2 - x_i^2}} \frac{x_i}{x} \right) \right]$$

$$(0 \le x \le L_1) \tag{40}$$

$$\int_{t_i}^{t} dt = n_e \sqrt{\frac{1}{wK_2}} \left[\sqrt{\lambda^2 - x^2} - \sqrt{\lambda^2 - x_i^2} - \lambda \ln \left(\frac{\lambda + \sqrt{\lambda^2 - x^2}}{\lambda + \sqrt{\lambda^2 - x_i^2}} \frac{x_i}{x} \right) \right]$$

$$(L_1 \le x \le x_{tr}) \tag{41}$$

$$\int_{t_i}^{t} dt = n_e \sqrt{\frac{\rho_s}{(\rho_s - \rho_f)wK_2}} \left[\sqrt{L^2 - x^2} - \sqrt{L^2 - x_i^2} - L \ln \left(\frac{L + \sqrt{L^2 - x^2}}{L + \sqrt{L^2 - x_i^2}} \frac{x_i}{x} \right) \right]$$

$$(x_{tr} \le x \le L_1 + L_2) \tag{42}$$

$$\text{where } \varphi = \sqrt{L_1^2 + \frac{K_1}{K_2}(2L_1 + L_2)L_2 + \frac{\rho_s K_1 H_0^2}{\rho_f w}}, \quad \lambda = \sqrt{(L_1 + L_2)^2 + \frac{\rho_s K_2 H_0^2}{\rho_f w}}.$$

Discussion of the analytical solutions

The changes of groundwater travel times due to the coastal reclamation are here discussed with a hypothetical example. Assume that the saturated hydraulic conductivity of the original aquifer is 3 m/day and the distance from the ground water divide to the original coastline is 1000m. The infiltration rate is 0.0005 m/day and H_0 =20m.

Figure 13 shows changes of the transit times from a point at which x = 800m to the sea with hydraulic conductivity of the reclamation materials for different reclamation lengths. The increase of the transit times due to reclamation decreases when K_2 increases and L_2 decreases. This is easily understood because the damping effect due to reclamation will be more significant when K_2 decreases and L_2 increases. With K_2 = 0.5 m/day the change is 29.7 years for L_2 = 500 m and 5.5 years for L_2 = 100 m, showing that changes of transit times are very sensitive to the reclamation length. Changes of the transit times tend to be low and insensitive to K_2 when K_2 is greater than 3m/day.

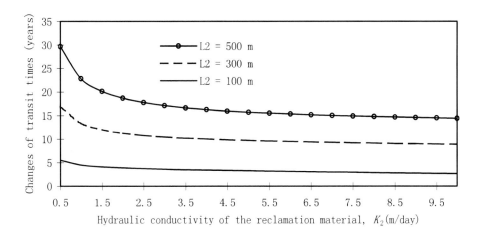

Figure 13. Changes of transit times from a point at which x = 800m to the sea with hydraulic conductivity of the reclamation material for different reclamation lengths when K_1 = 3 m/day, L_1 = 1000 m, ρ_s = 1.025 g/cm³, ρ_f = 1.000 g/cm³, w = 0.0005 m/day, and H_0 = 20m.

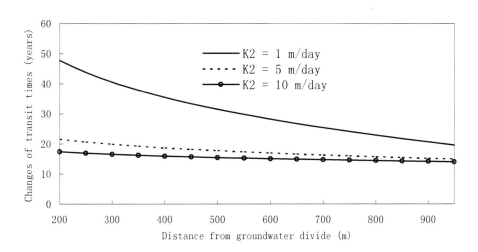

Figure 14. Changes of transit times to the sea from different positions in the original aquifer with distance from the water divide for different hydraulic conductivity of the reclamation material when K_1 = 3 m/day, L_1 = 1000 m, L_2=500m, ρ_s = 1.025 g/cm³, ρ_f = 1.000 g/cm³, w = 0.0005 m/day, and H_0 = 20m.

Figure 14 shows how transit times to the sea from different positions in the original aquifer change with the distance from the water divide for different hydraulic conductivity of the reclamation materials when L_2=500m. After reclamation, the water table in the whole original aquifer will increase so that the thickness of freshwater is increased and the water will flow more slowly. Therefore, as showed in Figure 14, the transit time changes increase from the original coastline to the water divide and tend to be more significant when K_2

becomes low. With $K_2 = 1$ m/day the change is 47.7 years for $x = 200$ m and 19.6 years for $x = 950$ m.

Figure 15 shows how ground water travel times from a point at which x = 800m to the original coastline change with hydraulic conductivity of the reclamation materials for different reclamation lengths. After reclamation groundwater flow in the original aquifer tends to be slower so that the travel time becomes greater. This is because the reclamation increases the water table so that the water will flow through a cross-sectional area of greater height. The influence of the reclamation on groundwater travel times in the original aquifer diminishes gradually when K_2 increases and L_2 decreases, and becomes less sensitive to K_2 when K_2 is greater than 7 m/day. With $L_2 = 500$ m the change is 7.7 years for $K_2 = 0.5$ m/day and 0.9 years for $K_2 = 10$ m/day.

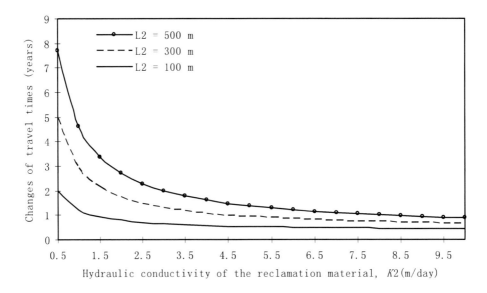

Figure 15. Changes of ground water travel times from a point at which x = 800m to the original coastline with different hydraulic conductivity of the reclamation materials for different reclamation lengths when $K_1 = 3$ m/day, $L_1 = 1000$ m, $\rho_s = 1.025$ g/cm^3, $\rho_f = 1.000$ g/cm^3, w = 0.0005 m/day, and $H_0 = 20$m.

Figure 16 shows the changes of ground water travel times from different positions in the original aquifer to the original coastline with distance from the water divide for different hydraulic conductivity of the reclamation materials when L_2=500m. Since the ground water in the whole original aquifer flows more slowly after reclamation, the increase of ground water travel times decreases with the distance from the ground water divide, which is similar to curves in Figure 14. When K_2 is low, the influence of the reclamation on groundwater flow will be more significant so that the changes of ground water travel times are greater. With $K_2 = 1$ m/day the change is 29.4 years for $x = 200$ m and 1.3 years for $x = 950$ m, indicating that changes of travel times are very sensitive to the distance towards the water divide.

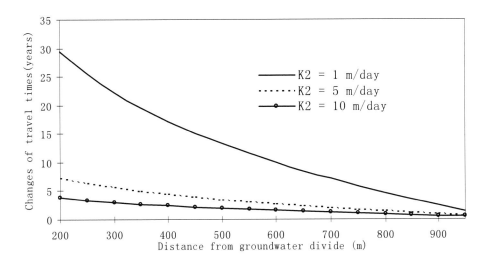

Figure 16. Changes of ground water travel times from different positions in the original aquifer to the original coastline with distance from the water divide for different hydraulic conductivity of the reclamation material when $K_1 = 3$ m/day, $L_1 = 1000$ m, $L_2 = 500$m, $\rho_s = 1.025$ g/cm^3, $\rho_f = 1.000$ g/cm^3, $w = 0.0005$ m/day, and $H_0 = 20$m.

2.3. Impact of Reclamation on Streamline Pattern and Flow Velocity

It is well known that recharge can cause the spreading of contaminants (e.g. plume diving), which should be considered when placing wells at sites. The streamlines originating at the locations for the source of contaminants can be used to decide how much spreading of contaminants to expect at a point down gradient from the source.

Streamline calculation
    The calculation method will be applicable for a case of one-dimensional shallow flow in a water table aquifer receiving an infiltration of a constant rate along the entire upper boundary (Figure 17). The base of the aquifer is assumed to be horizontal and impermeable. Water entering the aquifer due to recharge along the water table would move down gradient along the streamline. In a Dupuit-type water table aquifer recharged by surface infiltration and discharged by a down-gradient, fixed-head boundary, the flux is related to the recharge, w, by

$$Q_x = w(x\text{-}x_0) \tag{43}$$

where Q_x is the discharge per unit width (L^2/T), and x_0 is the location of the upgradient water divide. The total discharge in the x direction per unit width, through a vertical cross section of height $b(x)$ is partitioned by the streamline originating at $x = x_s$ (Strack ,1984), which leads to

$$\frac{Q_x^{t}}{Q_x} = \frac{D(x)}{b(x)} = \frac{Q_x - Q_s}{Q_x} = \frac{w(x - x_s)}{w(x - x_0)} = \frac{x - x_s}{x - x_0} \qquad (44)$$

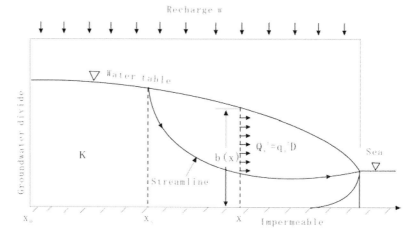

Figure 17. The streamline in an unconfined system recharged only by infiltration and discharged by a down-gradient, fixed-head boundary.

where Q_x^{t} is discharge above the streamline, Q_x is the total discharge, and $D(x)$ is the depth of the streamline below the water table. Note that (44) implies that Dupuit assumption applies, i.e., equipotential surfaces are vertical and the flow is essentially horizontal. In other words, the water head $h(x)$ and the specific discharge $q(x)$ are constant over the height of the aquifer. Thus the depth of the streamline below the water table, $D(x)$, can be obtained as:

$$D(x) = \frac{x - x_s}{x - x_0} b(x) \qquad (45)$$

Discussion of the analytical solutions

The streamlines can be calculated by combing Equations (30)-(31), (37) - (39), and (45). Assume that the saturated hydraulic conductivity K_1 of the aquifer is 0.1 m/day and the distance from the ground water divide to the original coastline, L_1, is 1000m. The densities of seawater and freshwater are 1.025 g/cm^3 and 1.000 g/cm^3, respectively. The infiltration rate is 0.0005 m/day and H_0 =20m. The hydraulic conductivity of the reclamation material, K_2, is 0.5 m/day and the reclamation length, L_2, is 500 m.

Figure 18 shows that how the water table and the streamline originating at x = 500m change before and after reclamation. One can see that the water table in the original aquifer increases significantly after reclamation. The streamline originating at x=500m also increases so that it may intersect the well (Figure 18). After reclamation the contaminants may enter some wells due to the increase of the elevation of the streamline originating from the source of contaminants, which should be considered when placing wells. One can also see that the recharged water will move across a vertical cross section of greater height so that the flow velocity becomes slower.

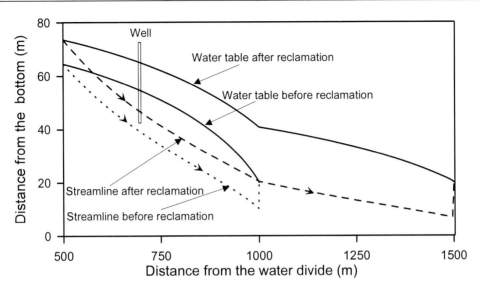

Figure 18. The water table and streamline originating at $x = 500$ m before and after reclamation, respectively (K_1=0.1m/day, K_2=0.5m/day, L_1=1000m, L_2=500m, ρ_s =1.025g/cm^3, ρ_f =1.000g/cm^3, w=0.0005m/day, and H_0=20m).

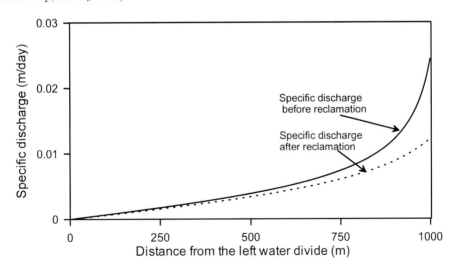

Figure 19. The specific discharge with distance from the water divide in the original aquifer before and after reclamation (K_1=0.1m/day, K_2=0.5m/day, L_1=1000m, L_2=500m, ρ_s =1.025g/cm^3, ρ_f =1.000g/cm^3, w=0.0005m/day, and H_0=20m).

Figure 19 shows how the specific discharge (Darcy flux) defined by Equation (25) changes before and after reclamation. The discharge per unit width will increase and the thickness of freshwater will decrease with distance from the water divide. Thus, as showed in Figure 19, the specific discharge increases with the distance from the water divide. After reclamation the height of the vertical cross section through which freshwater flows becomes greater because the reclamation increases the water table in the original aquifer. The discharge defined by Equation (43) remains unchanged so that the specific discharge decreases. As discussed previously, the maximal water level increase due to reclamation is at the original coastline, and the flow velocity at the water divide is always zero. Therefore, after reclamation the specific discharge becomes lower than that before reclamation especially in areas close to the original coastline.

3. IMPACT OF LAND RECLAMATION IN AN ISLAND

When the reclamation scale is relatively small compared to the size of the original groundwater catchment, the previous assumption that the water divide remains unchanged (Figure 1) is valid. In the situation of an island the water divide may be moved when large-scale land reclamation occurs. In this case the reclamation on one side may change the ground water flow on both sides of the island.

3.1. Impact of Reclamation on Ground Water Level and Saltwater Interface

Two kinds of islands are considered, both with ground water flow resulting from precipitation recharge. In these two different kinds of islands, the unconfined flow system is bounded at the bottom by a horizontal impermeable layer (Figure 20) and by a saltwater-freshwater interface (Figure 27), respectively.

3.1.1. An Island Bounded Below by a Horizontal Impermeable Layer

3.1.1.1. Solution for Ground Water Level and Saltwater Interface

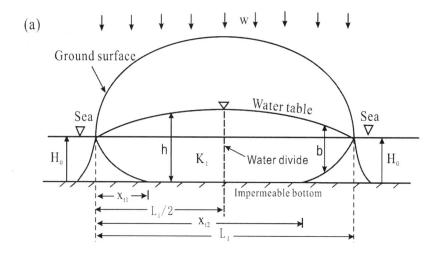

Figure 20(a).

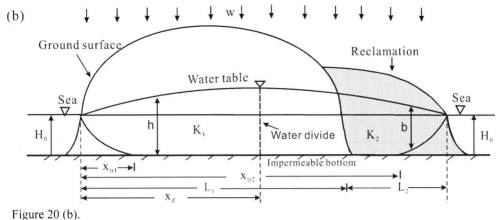

Figure 20 (b).

Figure 20. A schematic sketch of the ground water flow system in an island, which is bounded below by a horizontal impermeable layer and receives uniform vertical recharge (a) before reclamation and (b) after reclamation (Guo and Jiao, 2007).

Figure 20a shows an unconfined ground water system in a long island supported by a constant infiltration and bounded below by a horizontal impermeable layer. It can be a cross section through an elongated island, or a peninsula (Jiao et al. 2001). For convenience of discussion, it is referred to as a cross section in an island in the following discussion. The origin is located at the left coastline. L_1 is the width of the island before reclamation, and w is recharge rate per unit area. The distances from the left coastline to the tips of saltwater tongues are denoted as x_{t1} and x_{t2}, respectively. Guo and Jiao (2007) presented the solutions for the water levels which can be described as:

$$h = \sqrt{\frac{w(\rho_s - \rho_f)}{K_1 \rho_s}(L_1 x - x^2)} + H_0 \qquad (0 \leq x \leq x_{t1} \text{ or } x_{t2} \leq x \leq L_1)$$

$$(46)$$

$$h = \sqrt{\frac{w}{K_1}(L_1 x - x^2) + \frac{\rho_s}{\rho_f} H_0^2} \qquad (x_{t1} \leq x \leq x_{t2}) \qquad (47)$$

At the tip of the saltwater tongue, the water head is equal to $\rho_s H_0 / \rho_f$ (see Equation 4). Set h equal to $\rho_s H_0 / \rho_f$ in either (46) or (47), one has

$$x^2 - L_1 x + \frac{K_1(\rho_s^2 - \rho_s \rho_f)}{w \rho_f^2} H_0^2 = 0 \qquad (48)$$

The locations of the tips of saltwater tongues can be readily obtained by solving Equation (48) for x_{t1} and x_{t2}.

Figure 20b shows the unconfined ground water flow system in the island after reclamation. The hydraulic conductivity of the reclamation material and the reclamation length are denoted as K_2 and L_2, and the post-reclamation groundwater divide is located at $x = x_d$. After reclamation, the distances from the left coastline to the tips of saltwater tongues are represented as x_{tr1} and x_{tr2}, respectively. The solutions for the water level and the position of the ground water divide are (Guo and Jiao, 2007):

$$x_d = \frac{K_1 L^2 + (K_2 - K_1)L_1^2}{2(L_1 K_2 + L_2 K_1)} \tag{49}$$

$$h = \sqrt{\frac{w(\rho_s - \rho_f)}{K_1 \rho_s}(\beta x - x^2)} + H_0 \qquad\qquad 0 \le x \le x_{tr1} \tag{50}$$

$$h = \sqrt{\frac{w}{K_1}(\beta x - x^2) + \frac{\rho_s}{\rho_f} H_0^2} \qquad\qquad x_{tr1} \le x \le L_1 \tag{51}$$

$$h = \sqrt{\frac{w}{K_2}(\beta x - x^2 + 2\gamma / w) + \frac{\rho_s}{\rho_f} H_0^2} \qquad\qquad L_1 \le x \le x_{tr2} \tag{52}$$

$$h = \sqrt{\frac{w(\rho_s - \rho_f)}{K_2 \rho_s}(\beta x - x^2 + 2\gamma / w)} + H_0 \qquad\qquad x_{tr2} \le x \le L \tag{53}$$

where $\beta = \left[K_1 L^2 + (K_2 - K_1)L_1^2\right]/(L_1 K_2 + L_2 K_1)$ and $\gamma = w L_1 L_2 L (K_2 - K_1)/(2 L_1 K_2 + 2 L_2 K_1)$.

Set h equal to $\rho_s H_0 / \rho_f$ in either (50) or (51), one has

$$x^2 - \beta x + \frac{K_1(\rho_s^2 - \rho_s \rho_f)}{w \rho_f^2} H_0^2 = 0 \tag{54}$$

The distance from the left coastline to the tip of the saltwater tongue on the unreclaimed side, x_{tr1}, is one root of equation (54). Similarly, set h equal to $\rho_s H_0 / \rho_f$ in either (52) or (53), the location of the tip of the saltwater tongue on the reclaimed side, x_{tr2}, can be obtained as a root of the following equation:

$$x^2 - \beta x + \frac{K_2(\rho_s^2 - \rho_s \rho_f)}{w \rho_f^2} H_0^2 - 2\gamma / w = 0 \tag{55}$$

Solving (54) and (55) for the tip locations is not difficult, but the process is cumbersome and is not shown here.

An important influence of the reclamation on the ground water flow system is that the water divide is displaced. The displacement of the water divide, Δd, can be calculated as

$$\Delta d = x_d - \frac{L_1}{2} = \frac{K_1 L_2 (L_1 + L_2)}{2(L_1 K_2 + L_2 K_1)} \tag{56}$$

Equation (56) shows that the ground water divide will move toward the post-reclamation coastline, implying that ground water discharge to the sea on the left is increased. The displacement of the water divide is significant when K_1 is great. For a particular coastal area, the parameters L_1 and K_1 are fixed. In this case, Equation (56) implies that he displacement of the water divide, Δd, increases with L_2 and decrease with K_2. One can also see that the displacement of the water divide has nothing to do with the recharge rate w and the elevation of the sea level H_0.

3.1.1.2. Discussion of the Analytical Solutions Using a Hypothetical Example

A hypothetical example is employed to study the influence of the reclamation on ground water level and seawater-freshwater interface in an island bounded below by a horizontal impermeable layer. Assume that the width of the island before reclamation L_1 and the hydraulic conductivity of the original aquifer K_1 are 2000m and 0.1 m/day, respectively. The infiltration rate is taken as 0.0005 m/day and H_0 = 20m. The densities of seawater and freshwater are 1.025 g/cm^3 and 1.000 g/cm^3, respectively.

Change of the water level in the original aquifer

A comparison of predicted water levels at the original coastline on the reclaimed side (Figure 21) shows that the rise of the water level decreases with K_2 and increases with L_2. For a fixed value of L_2, the water level change tends to be sensitive to K_2 when K_2 is low. With L_2 = 500 m the change is 20 m for K_2 = 0.5 m/day and only 1.7 m for K_2 = 10 m/day, which indicates that the water level change can be very sensitive to K_2. The reclamation results in rise of the water table throughout the island (Figure 22), thereby increasing the volume of the freshwater. The water-table rise decreases with K_2 and increases with distance towards the left coastline. The shape of the water table in the original aquifer tends to be asymmetric after reclamation due to displacement of the water divide, as indicated by Equation (56). There is an obvious jump for the hydraulic gradient at the original coastline on the reclaimed side.

Predicted displacement of the tip of the saltwater tongue on the reclaimed side as a function of the hydraulic conductivity of the reclamation material and reclamation length is presented in Figure 23, which shows the displacement of the tip decreases with K_2 and increases with L_2. After reclamation, the saltwater interface on the reclamation side is pushed toward the post-reclamation coastline, which in turns increases the volume of the freshwater. The displacement of the tip is similar to the results in Figure 7. For the sake of convenience, it is not presented here the influence of the reclamation on the saltwater tongue on the unreclaimed side. After reclamation, the water divide moves toward the reclaimed side, increasing the ground water discharge to the sea on the unreclaimed side. Thus, the saltwater interface on the unreclaimed side will also move seaward. Therefore, for a large-scale land reclamation project, the changes of ground water flow and seawater-freshwater interface over

the whole island should be taken into account. Usually, reclamation will increase the volume
of the freshwater, a valuable water resource in the island situation.

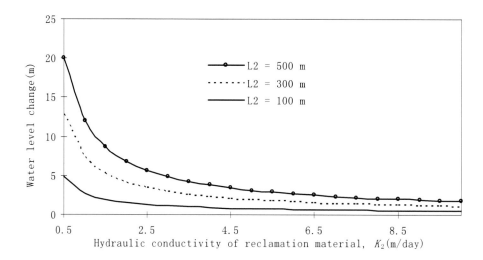

Figure 21. Change of the water level at the original coastline with hydraulic conductivity of the
reclamation material for different reclamation lengths when K_1=0.1m/day, L_1=2000m, ρ_s
=1.025g/cm^3, ρ_f =1.000g/cm^3, w=0.0005m/day, and H_0=20m.

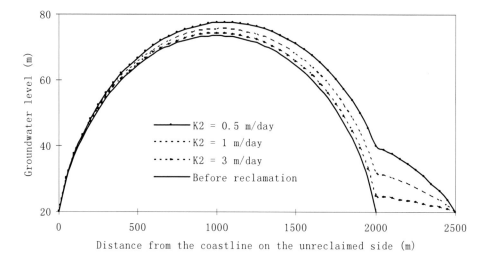

Figure 22. Water tables in the island with the distance from the coastline on the unreclaimed side for
different hydraulic conductivity of the reclamation material when K_1=0.1m/day, L_1=2000m, L_2=500m,
ρ_s =1.025g/cm^3, ρ_f =1.000g/cm^3, w=0.0005m/day, and H_0=20m (Guo and Jiao, 2007).

Displacement of the tip of the saltwater tongue

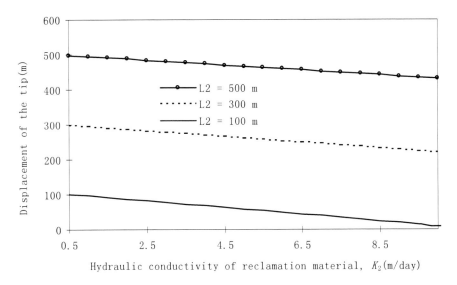

Figure 23. Displacement of the tip of the saltwater tongue on the reclaimed side with hydraulic conductivity of the reclamation material when the reclamation lengths are 500m, 300m and 100m, respectively (K_1=0.1m/day, L_1=2000m, ρ_s =1.025g/cm³, ρ_f =1.000g/cm³, w=0.0005m/day, and H_0=20m).

Change of the discharge to the sea on the unreclaimed side

Based on Equation (56), the percentage increase in flow to the sea on the unreclaimed side ($100 \times 2w\Delta d / wL_1 = 200\Delta d / L_1$) is plotted with the ratio of the reclamation length (L_2) to island width (L_1) (Figure 24). The percentage increase in flow increases with the ratio of the reclamation length to island width and decreases with K_2. With L_2 = 500 m, i.e., the ratio of the reclamation length to island width is 0.25, the percentage increase in flow to the sea on the unreclaimed side is 6% for K_2 = 0.5 m/day and 1% for K_2 = 3 m/day, indicating that the increase in discharge to the sea on the unclaimed side is very sensitive to K_2.

Figure 25 shows streamlines originating at x=500m and 1500m before and after reclamation. One can see that the elevations of streamlines on both sides of the island increase. On the unreclaimed side the streamline increases slightly more greatly than the water table so that the recharged water moves through a vertical cross section of lower height and the flow velocity of ground water becomes greater. On the reclaimed side, however, the water will flow more slowly because the increase of the streamline elevation is less than that of the water table.

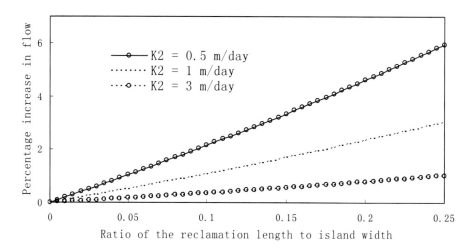

Figure 24. Percentage increase in the freshwater discharge to the unreclaimed side of the island versus the ratio of the reclamation length to island width for different K_2 when K_1=0.1m/day, L_1=2000m, ρ_s =1.025g/cm^3, ρ_f =1.000g/cm^3, w=0.0005m/day, and H_0=20m (Guo and Jiao, 2007).

Change of the streamline pattern and flow velocity

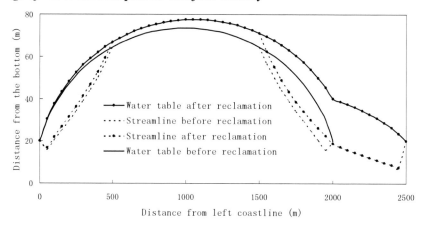

Figure 25. Change of the streamline pattern originating at x=500m and 1500m due to land reclamation with distance from the coastline on the unreclaimed side when K_1 = 0.1 m/day, K_2 = 0.5 m/day, L_1 = 2000 m, L_2 = 500 m, ρ_s = 1.025 g/cm^3, ρ_f = 1.000 g/cm^3, w = 0.0005 m/day, and H_0 = 20 m.

Figure 26 shows how the specific discharge and the water divide change in the original aquifer system before and after reclamation. Before reclamation the specific discharge decreases from the left coastline to the water divide and then increases gradually towards the coastline on the right. After reclamation, the specific discharge tends to increase on the unreclaimed side and decreases on the reclaimed side. The two curves intersect at a point in the area between the original water divide and the post-reclamation water divide. The influence of the reclamation on the flow velocity on the reclaimed side tends to be more significant than that on the unreclaimed side, and the influence increases with the distance from the left coastline.

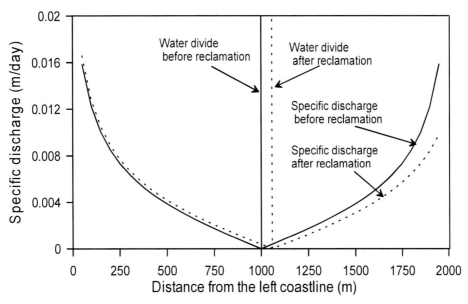

Figure 26. The specific discharge and the water divide in the original aquifer before and after reclamation (K_1=0.1m/day, K_2=0.5m/day, L_1=2000m, L_2=500m, ρ_s =1.025g/cm^3, ρ_f =1.000g/cm^3, w=0.0005m/day, and H_0=20m).

3.1.2. An Island Bounded Below by Saltwater-freshwater Interface

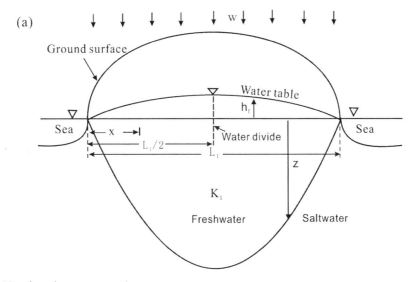

Figure 27. (Continued on next page.)

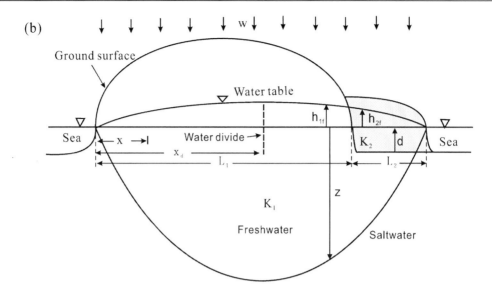

Figure 27. A schematic sketch of the ground water flow system beneath an oceanic island, which is bounded on the bottom by saltwater-freshwater interface and receives uniform vertical recharge : (a) before reclamation; (b) after reclamation.

3.1.2.1. Analytical Solution for Ground Water Level and Saltwater Interface

Figure 27a shows an approximate description of a long oceanic island supported by a constant rainfall infiltration rate. The freshwater lens, floating on top of sea water, is bounded above by a phreatic surface and below by a seawater-freshwater interface. The flow of the salt water beneath the interface may be neglected if the aquifer is thick. Here, x is the horizontal distance which origins at the left shoreline, L_1 is the original width of the island, w is the uniform vertical recharge rate per unit area, and K_1 is the hydraulic conductivity of the aquifer. The ocean surface is taken as the datum for the water head of fresh water $h_f(x)$, and $z(x)$ denotes the depth of the saltwater-freshwater interface below the sea level. The governing equation for freshwater flow based on Dupuit assumption and Ghyben-Herzberg relation is:

$$w(x - L_1/2) = -K_1(1+\delta)h_f \frac{dh_f}{dx} \qquad (57)$$

where $\delta = \rho_f /(\rho_s - \rho_f)$. The boundary conditions are $h_f = 0$ at $x = 0$ (or L_1) and $dh_f / dx = 0$ at $x = L_1/2$. The solution to equation (57) is

$$h_f = \sqrt{\frac{w}{K_1(1+\delta)}(L_1 x - x^2)} \qquad (58)$$

The expression (58) is identical to the solution presented by Henry (1964) and Fetter (1972).

Figure 27b shows the unconfined ground water flow system after reclamation. The reclamation length and the hydraulic conductivity of the reclamation material are denoted as

L_2 and K_2, respectively. The water depth of the shallow sea is d. The water divide will shift to the right due to reclamation because ground water cannot discharge to the sea on the right as freely as it was before reclamation. Assuming that the water divide after reclamation is at $x = x_d$, the mathematical model can be described as:

$$w(x - x_d) = -K_1(1 + \delta)h_{1f}\frac{dh_{1f}}{dx} \qquad (0 \le x \le L_1) \tag{59}$$

$$w(x - x_d) = -K_2(h_{2f} + d)\frac{dh_{2f}}{dx} - K_1(\delta h_{2f} - d)\frac{dh_{2f}}{dx} \quad (L_1 \le x \le L_1 + L_2) \tag{60}$$

with boundary conditions

$$h_{1f}(0) = 0 \tag{61}$$

$$h_{2f}(L_1 + L_2) = 0 \tag{62}$$

and the continuity conditions of ground water head and flux at the position where $x = L_1$ are

$$h_{1f}(L_1) = h_{2f}(L_1) \tag{63}$$

$$q_1(L_1) = q_2(L_1) \text{, or}$$

$$K_1(1 + \delta)h_{1f}\frac{dh_{1f}}{dx} = K_2(h_{2f} + d)\frac{dh_{2f}}{dx} + K_1(\delta h_{2f} - d)\frac{dh_{2f}}{dx} \tag{64}$$

where h_{1f} and h_{2f} are the water head of fresh water in the original aquifer and the reclamation area, respectively. The solution of Equations (59) and (61) is

$$h_{1f} = \sqrt{\frac{w}{K_1(1 + \delta)}(2x_d x - x^2)} \qquad (0 \le x \le L_1) \tag{65}$$

The solution of Equations (60) and (62) is

$$\frac{K_2 + \delta K_1}{2}h_{2f}{}^2 + d(K_2 - K_1)h_{2f} = $$

$$wx_d x - \frac{w}{2}x^2 + \frac{w}{2}(L_1 + L_2)^2 - wx_d(L_1 + L_2) \qquad (L_1 < x \le L_1 + L_2) \tag{66}$$

The following equation can be obtained by combining Equations (63), (64), (65) and (66):

$$\kappa t^2 + v t - (L_1 L_2 + L_1^2) = 0 \qquad (67)$$

where

$$t = \sqrt{\frac{w}{K_1(1+\delta)}(2x_d L_1 - L_1^2)}, \qquad (68)$$

$$\kappa = \frac{L_1 K_2 + L_2 K_1 + \delta K_1 (L_1 + L_2)}{w L_2} \qquad (69)$$

$$v = \frac{2d L_1 (K_2 - K_1)}{w L_2} \qquad (70)$$

The solution to Equations (67) through (70) is

$$x_d = \frac{L_1}{2} + \frac{K_1(1+\delta)t^2}{2w L_1} \qquad (71)$$

$$t = \frac{\sqrt{v^2 + 4\kappa(L_1 L_2 + L_1^2)} - v}{2\kappa} \qquad (72)$$

The solutions for ground water level and the seawater-freshwater interface after reclamation can be obtained with the combination of Equations (65), (66), (71) and (72). It should be noted that the ocean level is not involved in all these equations, indicating that the ocean level does not impact the shape and, thus, the volume of freshwater. However, as discussed previously, the volume of freshwater lens can be greatly impacted by the sea level if the island is bounded below by a horizontal impermeable bottom.

Equation (71) shows the water divide moves towards the coastline on the reclaimed side after reclamation, the change in the water divide is calculated as

$$\Delta d = x_d - \frac{L_1}{2} = \frac{K_1(1+\delta)t^2}{2w L_1} \qquad (73)$$

The increased freshwater discharge to the sea on the unreclaimed side is calculated as:

$$w\Delta d = \frac{K_1(1+\delta)t^2}{2L_1} \qquad (74)$$

3.1.2.2. Discussion of the Analytical Solutions Using a Hypothetical Example

A hypothetical example is employed to study the influence of the reclamation on the ground water flow pattern and the seawater-freshwater interface. Assume that the distance between the coastlines before reclamation L_1 and the hydraulic conductivity of the original aquifer K_1 are 2000m and 0.1 m/day, respectively. The infiltration rate is taken as 0.0005 m/day. The densities of seawater and freshwater are 1.025 g/cm^3 and 1.000 g/cm^3, respectively.

Change of the water level at the original coastline and displacement of the water divide
Figure 28 shows how the water level at the original coastline changes with K_2 for different reclamation lengths when the reclamation depth d is 20m. The water-level rise decreases with K_2 and increases with L_2. The water level buildup is very sensitive to the reclamation length and the hydraulic conductivity of the reclamation material. The increase of the water table is 9.3 m with $K_2 = 0.5$ m/day and $L_2 = 500$ m, and is only 0.3 m with $K_2 = 10$ m/day and $L_2 = 100$m. Comparing the results in Figure 28 with those in Figure 21, one can see that the influence of reclamation on water-level change in the original aquifer in an island bounded below by a horizontal impermeable layer is more significant than that in an island bounded below by seawater-freshwater interface.

Based on Equation (73), the displacement of the water divide (Δd) with hydraulic conductivity of the reclamation material is showed in Figure 29. The displacement becomes smaller as K_2 increases and L_2 decreases. The curves show that the displacement becomes more sensitive to L_2 and K_2 when K_2 becomes low, and the displacement of the water divide can be very significant. With $K_2 = 0.5$ m/day and $L_2 = 500$ m, the displacement of the water divide is 176.8 m, a great distance relative to the width of 2000 m of the island before reclamation.

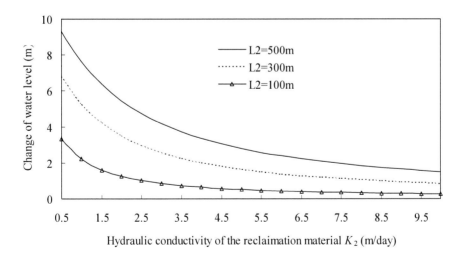

Figure 28. Change of the water level at the original coastline with the hydraulic conductivity of the reclamation material for different reclamation lengths when K_1=0.1m/day, L_1=2000m, ρ_s =1.025g/cm^3, ρ_f =1.000g/cm^3, w=0.0005 m/day, and d=20m.

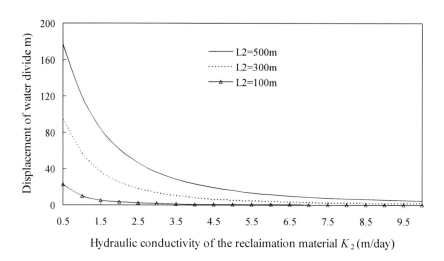

Figure 29. Displacement of ground water divide with hydraulic conductivity of the reclamation material for different reclamation lengths when K_1=0.1m/day, L_1=2000m, ρ_s =1.025g/cm^3, ρ_f =1.000g/cm^3, w=0.0005 m/day, and d=20m.

Change of the seawater-freshwater interface and the freshwater resource

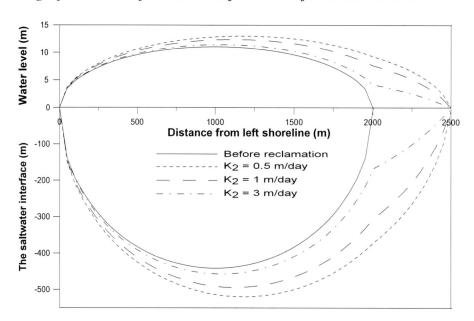

Figure 30. Changes of the water table and seawater-freshwater interface with distance from the coastline on the unreclaimed side when K_1 = 0.1 m/day, L_1 = 2000 m, L_2 = 500m ρ_s = 1.025g/cm^3, ρ_f = 1.000g/cm^3, w = 0.0005 m/day, and d = 20m.

Figure 30 shows changes of the water table (above the horizontal axis) and the seawater-freshwater interface (below the horizontal axis) with distance from the coastline on the unreclaimed side when L_2 and d are 500m and 20m, respectively. After reclamation, both the water table and the depth of the seawater-freshwater interface increase, thereby increasing the volume of the freshwater lens, an important resource in coastal areas. The increases in water table and the depth of the interface are more significant when K_2 decreases. The hydraulic gradient jumps at the interface between the original aquifer and the reclamation area due to the permeability contrast between the original aquifer and the reclamation material. When K_2 becomes greater, the permeability contrast will increase accordingly, thereby increasing the jump of the hydraulic gradient at the original coastline.

Figure 31 shows how the water table and the seawater-freshwater interface change with distance from the coastline on the unreclaimed side when L_2 and K_2 are 500m and 1m/day, respectively. The water table and the depth of the seawater-freshwater interface increase when the reclamation depth d decreases because K_2 (1m/day) is greater than K_1 (0.1m/day). If K_2 is lower than K_1, the water table and the depth of the seawater–freshwater interface can be expected to increase with the reclamation depth d.

For simplicity, it is not presented here how the water table and the seawater-freshwater interface change with distance from the left coastline for different reclamation lengths. The previous analysis (Figure 28) shows that the water table and the depth of the seawater-freshwater interface will increase when the reclamation length L_2 becomes great. The volume of the freshwater lens (Figure 27b), therefore, is increased.

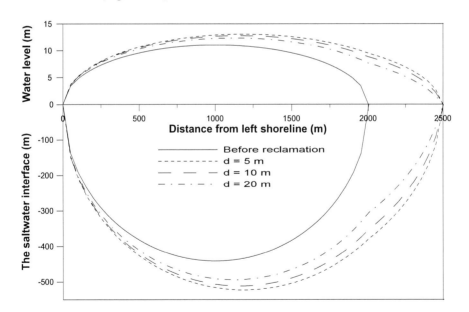

Figure 31. Change of the water table and the seawater-freshwater interface with distance from the coastline on the unreclaimed side when $K_1 = 0.1$ m/day, $K_2 = 1$ m/day, $L_1 = 2000$ m, $L_2 = 500$m $\rho_s = 1.025$g/cm^3, $\rho_f = 1.000$g/cm^3, $w = 0.0005$ m/day.

Change of the discharge to the sea on the unreclaimed side

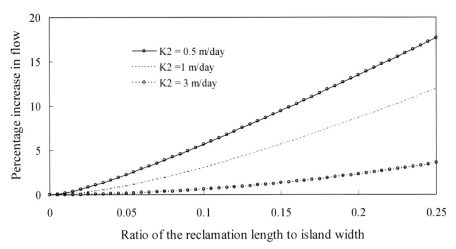

Figure 32. Percentage increase in the freshwater discharge to the unreclaimed side of the island versus the ratio of the reclamation length to island width for different K_2 when K_1=0.1m/day, L_1=2000m, ρ_s =1.025g/cm^3, ρ_f =1.000g/cm^3, w=0.0005 m/day, and d=20m.

Equation (74) shows that the freshwater recharge to the sea on the unreclaimed side will be increased. Using Equation (74), the percentage increase in the flow to the sea on the unreclaimed side ($100 \times 2w\Delta d / wL_1 = 200\Delta d / L_1$) is plotted with the ratio of the reclamation length (L_2) to island width (L_1) (Figure 32). The percentage increase in flow increases with the ratio of the reclamation length to island width and decreases with K_2, which is similar to the results in an island bounded below by an impermeable layer (Figure 24). When ratio of the reclamation length to island width is low and K_2 is great, the increase in flow to the left sea is minor and insensitive to K_2. The increase in flow, however, tends to be significant and sensitive to K_2 when ratio of the reclamation length to island width is great and K_2 is low, and increases almost linearly with ratio of L_2 to L_1. Comparing the results in Figure 32 with those in Figure 24, one can see that the increase in discharge to the sea on the unreclaimed side, and thus the displacement of the water divide (see Equation 74), is more significant in an island bounded below by seawater-freshwater interface than that in an island bounded below by a horizontal impermeable layer. With L_2 = 500 m, i.e., the ratio of the reclamation length to island width is 0.25, the percentage increase in flow to the sea on the unreclaimed side is 6% in an island bounded below by a horizontal impermeable layer, and 17% in an island bounded below by seawater-freshwater interface.

Change of the streamline pattern and the flow velocity

Figure 33 shows the streamlines originating at x=500m and 1500m in the original aquifer before and after reclamation. After reclamation the elevation of the streamline on the unreclaimed side becomes higher but that on the reclaimed side becomes lower. The post-reclamation water table on the unreclaimed side will increase, but the rise of the water table is less than that of the streamline elevation (Figure 33). Therefore, the infiltrated water will move across a vertical cross section of lower height, which leads to a greater flow velocity.

Similarly, the water will flow through a vertical cross section of greater height and the flow velocity becomes slower on the reclaimed side.

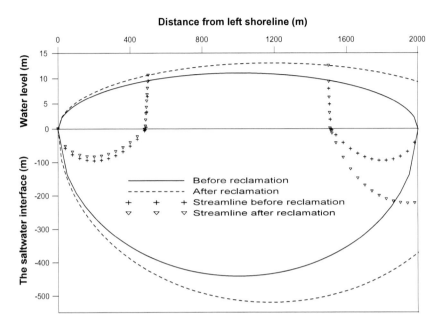

Figure 33. Change of streamlines originating at x=500m and 1500m in the original aquifer due to land reclamation with distance from the left shoreline when $K_1 = 0.1$ m/day, $K_2 = 0.5$ m/day, $L_1 = 2000$ m, $L_2 = 500$m $\rho_s = 1.025$g/cm^3, $\rho_f = 1.000$g/cm^3, $w = 0.0005$ m/day and $d = 20$ m.

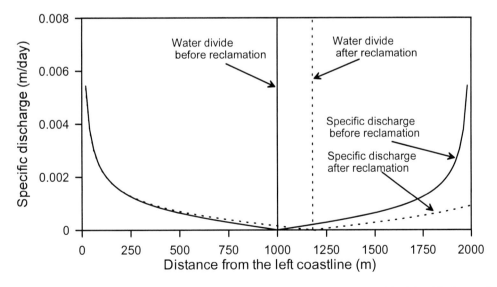

Figure 34. The specific discharge and the water divide in the original aquifer of the island before and after reclamation when $K_1 = 0.1$ m/day, $K_2 = 0.5$ m/day, $L_1 = 2000$ m, $L_2 = 500$m $\rho_s = 1.025$g/cm^3, $\rho_f = 1.000$g/cm^3, $w = 0.0005$ m/day and $d = 20$ m.

Figure 34 shows how the specific discharge and the water divide change in the original aquifer before and after reclamation. The specific discharge decreases from the coastline on the unreclaimed side to the water divide and then increases gradually towards the coastline on the right, which is similar to that in the island bounded below by an impermeable bottom (Figure 26). With the same parameter setting, the specific discharge (Figure 34) is much less than that in an island with a horizontal impermeable bottom. This is because the freshwater thickness is overall bigger so that the available surface for flow per unit width of aquifer is greater in an island bounded below by saltwater interface. After reclamation, the specific discharge tends to increase on the unreclaimed side and decrease on the reclaimed side, and the water divide moves towards the coastline on the reclaimed side. The displacement of the water divide is more significant than that in an island with a horizontal impermeable bottom (Figure 26) for the same parameter setting.

3.2. Impact of Land Reclamation on Ground Water Travel Time

In this section, analytical solutions are developed to estimate ground water travel times in unconfined aquifers located in an island bounded on the bottom by a horizontal impermeable layer, and the influence of the reclamation on ground water travel time is presented.

3.2.1. Analytical Solution

Before reclamation, Equation (34) remains valid in the portion between the water divide and the tip of the saltwater tongue. Equation (35) remains valid in the portion between the tip of the saltwater tongue and the corresponding coastline. In the coordinate system of the island setting, however, L_1 in Equations (34) and (35) should be replaced by $L_1/2$, x by $x - L_1/2$, and x_i by $x_i - L_1/2$. The solutions are:

$$\int_{t_i}^{t} dt = n_e \sqrt{\frac{\rho_s}{(\rho_s - \rho_f)wK_1}} \left[\sqrt{L_1 x - x^2} - \sqrt{L_1 x_i - x_i^2} - \frac{L_1}{2} \ln\left(\frac{L_1/2 + \sqrt{L_1 x - x^2}}{L_1/2 + \sqrt{L_1 x_i - x_i^2}} \frac{x_i - L_1/2}{x - L_1/2} \right) \right]$$

$$(0 \leq x \leq x_{t1} \quad or \quad x_{t2} \leq x \leq L_1) \tag{75}$$

$$\int_{t_i}^{t} dt = n_e \sqrt{\frac{1}{wK_1}} \left[\sqrt{\chi^2 - (x - L_1/2)^2} - \sqrt{\chi^2 - (x_i - L_1/2)^2} - \chi \ln\left(\frac{\chi + \sqrt{\chi^2 - (x - L_1/2)^2}}{\chi + \sqrt{\chi^2 - (x_i - L_1/2)^2}} \frac{x_i - L_1/2}{x - L_1/2} \right) \right]$$

$$(x_{t1} \leq x \leq x_{t2}) \tag{76}$$

where $\chi = \sqrt{L_1^2/4 + \dfrac{\rho_s K_1}{\rho_f w} H_0^2}$

Integrating Equations (75) and (76) with respect to x from x_0 to x_1 yields the travel time of water flow from one position ($x = x_0$) to another position ($x = x_1$) before reclamation. If $x_1 = 0$ (or L_1), the obtained travel time is the transit time of a particle of water originating at $x = x_0$.

After reclamation, Equation (34) remains valid in the portion between the water divide and the tip of the saltwater tongue on the unreclaimed side but with x_d in place of L_1 and $x-x_d$ in place of x in the coordinate system of the island setting. Equation (35) remains valid in the portion between the tip of the saltwater tongue on the unreclaimed side and the corresponding coastline and the same substitutions of symbols are made. Similarly, Equation (40) remains valid in the portion between the water divide and the original coastline on the reclaimed side, Equation (41) remains valid between the original coastline and the tip of the saltwater tongue on the reclaimed side, and Equation (42) remains valid between the tip of the saltwater tongue and the coastline on the reclaimed side. However, the symbol L_1 should be replaced by $L_1 - x_d$ and x replaced by $x-x_d$. The final solutions are:

$$\int_{t_i}^{t} dt = n_e \sqrt{\frac{\rho_s}{(\rho_s - \rho_f)wK_1}} \left[\sqrt{2x_d x - x^2} - \sqrt{2x_d x_i - x_i^2} - x_d \ln\left(\frac{x_d + \sqrt{2x_d x - x^2}}{x_d + \sqrt{2x_d x_i - x_i^2}} \frac{x_i - x_d}{x - x_d} \right) \right]$$

$$(0 \le x \le x_{tr1}) \tag{77}$$

$$\int_{t_i}^{t} dt = n_e \sqrt{\frac{1}{wK_1}} \left[\sqrt{\varepsilon^2 - (x - x_d)^2} - \sqrt{\varepsilon^2 - (x_i - x_d)^2} - \varepsilon \ln\left(\frac{\varepsilon + \sqrt{\varepsilon^2 - (x - x_d)^2}}{\varepsilon + \sqrt{\varepsilon^2 - (x_i - x_d)^2}} \frac{x_i - x_d}{x - x_d} \right) \right]$$

$$(x_{tr1} \le x \le L_1) \tag{78}$$

$$\int_{t_i}^{t} dt = n_e \sqrt{\frac{1}{wK_2}} \left[\sqrt{\eta^2 - (x - x_d)^2} - \sqrt{\eta^2 - (x_i - x_d)^2} - \eta \ln\left(\frac{\eta + \sqrt{\eta^2 - (x - x_d)^2}}{\eta + \sqrt{\eta^2 - (x_i - x_d)^2}} \frac{x_i - x_d}{x - x_d} \right) \right]$$

$$(L_1 \le x \le x_{tr2}) \tag{79}$$

$$\int_{t_i}^{t} dt = n_e \sqrt{\frac{\rho_s}{(\rho_s - \rho_f)wK_2}} \left[\sqrt{(L - x_d)^2 - (x - x_d)^2} - \sqrt{(L - x_d)^2 - (x_i - x_d)^2} \right.$$

$$\left. -(L - x_d)\ln\left(\frac{L - x_d + \sqrt{(L - x_d)^2 - (x - x_d)^2}}{L - x_d + \sqrt{(L - x_d)^2 - (x_i - x_d)^2}} \frac{x_i - x_d}{x - x_d} \right) \right]$$

$$(x_{tr2} \le x \le L) \tag{80}$$

Where $x_d = \left[K_1 L^2 + (K_2 - K_1)L_1^2 \right] / \left[2(L_1 K_2 + L_2 K_1) \right]$,

$$\varepsilon = \sqrt{x_d^2 + \frac{\rho_s K_1}{\rho_f w} H_0^2} \text{ , and } \eta = \sqrt{(L_1 + L_2 - x_d)^2 + \frac{\rho_s K_2 H_0^2}{\rho_f w}} .$$

Integrating Equations (77), (78), (79) and (80) with respect to x from x_0 to x_1 yields the travel time of water flow from one position ($x = x_0$) to another position ($x = x_1$) in the unconfined aquifer after reclamation. Note that the water will flow towards the coastline on the reclaimed side when $x_0 > x_d$, and towards the coastline on the unreclaimed side when $x_0 < x_d$.

3.2.2. Discussion of the Analytical Solutions

Changes of ground water travel times due to the coastal reclamation are here discussed with a hypothetical example. Assume that the saturated hydraulic conductivity of the original aquifer is 3 m/day and the distance between the coastlines before reclamation, L_1, is 2000m. The infiltration rate is 0.0005 m/day and $H_0 =20$m. The effective porosity is assumed to be 0.3.

The influence of reclamation on travel time on the unreclaimed side

Figure 35 shows changes of the transit times from a point at x = 200m to the sea on the unreclaimed side with hydraulic conductivity of the reclamation material for different reclamation lengths. One can see that values of changes of the transit times are negative, which indicates the water flows faster after reclamation. The change of transit times decreases with K_2 and increases with L_2 because the influence of the reclamation becomes more significant when K_2 decreases or L_2 increases. The decrease of the travel time is 3.2 years with $K_2 = 0.5$ m/day and $L_2 = 500$ m, and is only 0.12 years with $K_2 = 10$ m/day and $L_2 = 100$m. Change of the transit time tends to be low and insensitive to K_2 when K_2 is greater than 6m/day. Figure 36 shows how transit times to the sea from different positions in the original aquifer change with distance from the left coastline for different hydraulic conductivity of the reclamation material when $L_2=500$m. Changes of transit times increase with the distance from the left coastline, and tend to be more significant when K_2 is low. With $x = 800$m, the change is 31 years for $K_2 = 1$ m/day and only 9.9 years for $K_2 = 10$ m/day.

The influence of reclamation on travel time on the reclaimed side

Figure 37 shows changes of travel times from a point at x = 1800 m to the original coastline on the reclaimed side with hydraulic conductivity of the reclamation material for different reclamation lengths. One can see that values of changes of travel times are positive, which indicates that the water flows more slowly after reclamation. The increase of the travel time decreases with K_2 and increases with L_2. Changes of travel times tend to be low and insensitive to K_2 when K_2 is greater than 5m/day. Comparision of Figures 35 and 37 shows that the influence of reclamation on ground water travel time is much more significant on the reclaimed side than that on the unreclaimed side. With $K_2 = 0.5$m/day and $L_2 = 500$ m, change of travel time over the last 200 m of the original aquifer is 72.8 years on the reclaimed side, and only 3.2 years on the unreclaimed side

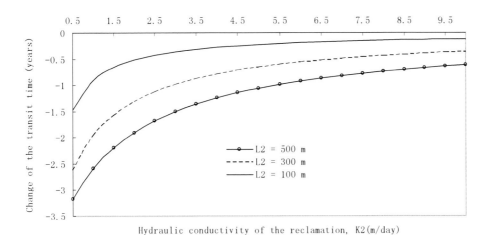

Figure 35. Changes of transit times from a point at x = 200m to the sea on the unreclaimed side with hydraulic conductivity of the reclamation material for different reclamation lengths when K_1 = 3 m/day, L_1= 2000 m, ρ_s = 1.025g/cm^3, ρ_f = 1.000g/cm^3, w = 0.0005 m/day and H_0=20m.

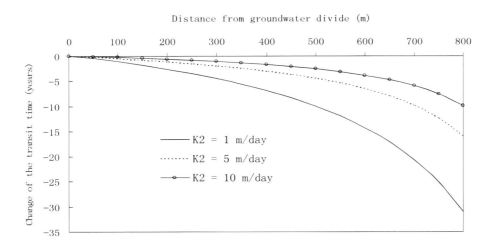

Figure 36. Changes of transit times to the sea on the unreclaimed side from different positions in the original aquifer with distance from the left coastline for different hydraulic conductivity of the reclamation materials when K_1= 3 m/day, L_1= 2000 m, L_2=500m ρ_s = 1.025g/cm^3, ρ_f = 1.000g/cm^3, w = 0.0005 m/day and H_0=20m.

Figure 38 shows how transit times from a point at x = 1800m to the sea on the reclaimed side change with hydraulic conductivity of the reclamation materials for different reclamation lengths. Ground water flow in the original aquifer tends to be slower so that the travel time becomes longer after reclamation (Figure 37). Also, the travel path of ground water becomes longer due to land reclamation. Therefore, transit times become greater than those before reclamation. The influence of the reclamation on groundwater transit time diminishes gradually when K_2 increases and L_2 decreases. For a fixed reclamation length, the impact of reclamation becomes less sensitive to K_2 when K_2 becomes great. Chang of transit times from a point at x = 1800 m can range from several years to more than one hundred years when K_2 and L_2 are combined differently. With K_2 = 0.5 m/day the change is 10.4 years for L_2 = 100 m and 123 years for L_2 = 500 m.

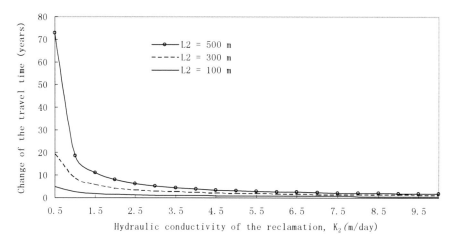

Figure 37. Changes of travel times from a point at x = 1800m to original coastline on the reclaimed side with hydraulic conductivity of the reclamation material for different reclamation lengths when K_1 = 3 m/day, L_1 = 2000 m, ρ_s = 1.025g/cm^3, ρ_f = 1.000g/cm^3, w = 0.0005 m/day and H_0 = 20 m.

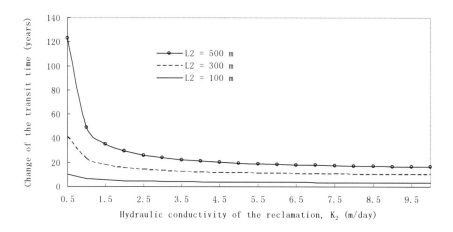

Figure 38. Changes of transit times from a point at x = 1800 m to the sea on the reclaimed side with hydraulic conductivity of the reclamation materials for different reclamation lengths when K_1 = 3 m/day, L_1 = 2000 m, ρ_s = 1.025g/cm^3, ρ_f = 1.000g/cm^3, w = 0.0005 m/day and H_0 = 20m.

4. SUMMARY AND CONCLUSIONS

The impact of land reclamation on ground water level and saltwater interface in an extensive land mass and an island are discussed. After reclamation, the water level in the original aquifer is increased and the saltwater interface moves seaward. The recharge can occur in a greater area so that the size of aquifer, and therefore the water resource, is increased. In the island situation, large-scale reclamation can result in the displacement of the water divide towards the reclaimed side, thereby increasing ground water discharge to the sea on both sides of the island. In all the cases, the impact of the reclamation on the water level and the saltwater interface mainly depends on the reclamation scale and the hydraulic conductivity of the reclamation material. In the case of an extensive land mass, the analytical solutions are checked against the results of a numerical code, which shows that only minor difference occurs.

Analytical solutions have been obtained to describe ground water travel time before and after reclamation for an extensive land mass and an island bounded below by a horizontal impermeable layer. Comparisons have been made between the solution in this paper and that developed by Chesnaux et al. (2005), which doesn't include the influence of seawater-freshwater interface. It is concluded that difference of these two solutions depends mainly on the elevation of the sea level, the infiltration rate and the hydraulic conductivity of the aquifer. The impact of reclamation on groundwater travel time is discussed detailedly based on the hypothetical examples. The groundwater travel time tends to be longer, and water flows more slowly in the original aquifer after reclamation. In the island situation, the ground water will flow more slowly in the original aquifer on the reclaimed side. However, the ground water travel time becomes shorter and water flows faster in the unreclaimed side. The impact of the coastal reclamation on ground water travel time is mainly controlled by the reclamation scale and hydraulic conductivity of the reclamation material. For the same reclamation scale and hydraulic conductivity of the reclamation material, the reclamation tends to have more significant impact on ground water flow on the reclaimed side than that on the unreclaimed side.

The impact of reclamation on flow velocity and streamline pattern is also presented. In some cases, the elevation of streamline increases after reclamation, which may impact spreading of contaminants and should be noted when placing wells. After reclamation the flow velocity will decrease especially in areas close to the original coastline for the situation of an extensive landmass. For the island setting, the flow velocity will decrease on the reclaimed side and increase on the unreclaimed side.

The analytical study in this paper is based on Dupuit and Ghyben-Herzberg assumptions. There exist some shortcomings in these assumptions such as no seepage face at the coastline and neglect of the vertical flow. In addition, the conceptual models considered here don not include the heterogeneity of the aquifer system, complicated topography at the coast, and the transient process immediately after the reclamation (Guo and Jiao, 2007). However, this study presents very useful insight into change of coastal ground water system in response to land reclamation.

APPENDIX A

A single-potential solution for regional interface problems in coastal aquifers was presented by Strack (1976), which was restricted to cases of steady flow with homogeneous isotropic permeability where the Dupuit assumption was valid. This approach is a more advanced and general method to handle combined unconfined flow, unconfined interface flow. In this

Appendix, we describe the derivation of the analytical solution for groundwater level and seawater-freshwater interface in an extensive land mass in terms of the potential function instead of water head, and all symbols used are as defined previously.

Solutions for water level and saltwater interface before reclamation

Figure 1a shows a coastal unconfined ground water system receiving uniform vertical recharge. According to Strack (1976), the potentials for combined unconfined flow and unconfined interface flow are

$$\Phi = \frac{1}{2} K_1 h^2 + C_u \qquad\qquad h \geq \frac{\rho_s}{\rho_f} H_0 \qquad\qquad (A1)$$

$$\Phi = \frac{1}{2} K_1 \frac{\rho_s}{\rho_s - \rho_f} (h - H_0)^2 + C_{ui} \qquad\qquad h \leq \frac{\rho_s}{\rho_f} H_0 \qquad\qquad (A2)$$

$$C_{ui} - C_u = \frac{1}{2} K_1 \frac{\rho_s}{\rho_f} H_0^{\;2} \qquad\qquad (A3)$$

where ρ_f and ρ_s are the densities of fresh ground water and seawater. C_{ui} and C_u stand for two constants and the subscripts ui and u refer to unconfined interface flow and unconfined flow, respectively. Choosing $C_u = 0$, one obtains

$$C_u = 0 \qquad\qquad C_{ui} = \frac{1}{2} K_1 \frac{\rho_s}{\rho_f} H_0^{\;2} \qquad\qquad (A4)$$

The boundary conditions of the unconfined aquifer before reclamation (Figure 1a) are

$$x = 0 \qquad\qquad Q_x = -\frac{\partial \Phi}{\partial x} = 0 \qquad\qquad (A5)$$

and

$$x = L_1 \qquad h = H_0 \qquad \Phi = \frac{1}{2} K_1 \frac{\rho_s}{\rho_f} H_0^{\;2} \qquad\qquad (A6)$$

where Q_x is the discharge vector defined as pointing in the direction of flow and having a magnitude equal to the discharge flowing through a surface perpendicular to the direction of flow, of unit width and of a height equal to the thickness of the flow region (Strack, 1976). The governing differential equation for the case of rainfall is

$$\nabla^2 \Phi = -w \qquad\qquad (A7)$$

The solution of Equations (A5) through (A7) is

$$\Phi = \frac{1}{2} w(L_1^2 - x^2) + \frac{1}{2} K_1 \frac{\rho_s}{\rho_f} H_0^2 \qquad (A8)$$

The heads in the aquifer can be obtained from (A1), (A2), and (A8):

$$h = \sqrt{\frac{w}{K_1}(L_1^2 - x^2) + \frac{\rho_s}{\rho_f} H_0^2} \qquad 0 \le x \le x_t \qquad (A9)$$

$$h = \sqrt{\frac{w(\rho_s - \rho_f)}{K_1 \rho_s}(L_1^2 - x^2) + H_0} \qquad x_t \le x \le L_1 \qquad (A10)$$

The expression for the location of the tip of the saltwater tongue can be obtained by setting h equal to $\rho_s H_0 / \rho_f$ in either (A1) or (A2) and then solving for x after substitution of this value of the potential in (A8), which yields

$$x_t = \sqrt{L_1^2 - \frac{K_1(\rho_s^2 - \rho_s \rho_f)}{w \rho_f^2} H_0^2} \qquad (A11)$$

Solutions for water level and saltwater interface after reclamation
After reclamation (Figure 1b), the potentials in the reclaimed portion of the aquifer can be defined as

$$\Phi = \frac{1}{2} K_2 h^2 \qquad\qquad h \ge \frac{\rho_s}{\rho_f} H_0 \qquad (A12)$$

$$\Phi = \frac{1}{2} K_2 \frac{\rho_s}{\rho_s - \rho_f}(h - H_0)^2 + \frac{1}{2} K_2 \frac{\rho_s}{\rho_f} H_0^2 \qquad h \le \frac{\rho_s}{\rho_f} H_0 \qquad (A13)$$

The condition (A5) remains valid for the left boundary but the boundary condition at $x = L = L_1 + L_2$, is changed as:

$$x = L \qquad h = H_0 \qquad \Phi = \frac{1}{2} K_2 \frac{\rho_s}{\rho_f} H_0^2 \qquad (A14)$$

The governing equation for the potential throughout the flow region is the same as (A7). The solution is

$$\Phi = \frac{1}{2} w(L^2 - x^2) + \frac{1}{2} K_2 \frac{\rho_s}{\rho_f} H_0^2 \qquad (A15)$$

If the hydraulic conductivity jumps at $x = L_1$ then (A15) remains valid, but the potential jumps:

$$\Phi^- / \Phi^+ = K_1 / K_2 \qquad (A16)$$

In order to have a continuous potential, the potential in the original aquifer is redefined as

$$\Phi = \frac{1}{2} K_1 h^2 + \xi \qquad (A17)$$

where $\xi = \Phi^+ - \Phi^- = \dfrac{K_2 - K_1}{K_2} \Phi^+$.

Now that the new potential is continuous throughout the flow region. The unknown value ξ can be obtained from (A15) by computing Φ at $x = L_1$:

$$\xi = \frac{K_2 - K_1}{K_2} \Phi(L_1) = \frac{K_2 - K_1}{2K_2} \left[w(L^2 - L_1^2) + K_2 \frac{\rho_s}{\rho_f} H_0^2 \right] \qquad (A18)$$

The solution throughout the flow domain is now obtained. The head in the original aquifer can be computed from (A15) and (A17), and heads in the reclaimed portion of the aquifer can be obtained from (A12), (A13), and (A15):

$$h = \sqrt{ w \left[\frac{1}{K_1}(L_1^2 - x^2) + \frac{1}{K_2}(2L_1 + L_2)L_2 \right] + \frac{\rho_s}{\rho_f} H_0^2 } \qquad 0 \le x \le L_1 \qquad (A19)$$

$$h = \sqrt{ \frac{w}{K_2} \left[(L_1 + L_2)^2 - x^2 \right] + \frac{\rho_s}{\rho_f} H_0^2 } \qquad L_1 \le x \le x_{tr} \qquad (A20)$$

$$h = \sqrt{\frac{w(\rho_s - \rho_f)}{K_2 \rho_s}\left[(L_1 + L_2)^2 - x^2\right]} + H_0 \qquad x_{tr} \leq x \leq L_1 + L_2$$

$$(A21)$$

The expression for the location of the tip of the saltwater tongue can be obtained by setting h equal to $\rho_s H_0 / \rho_f$ in either (A12) or (A13) and then solve for x after substitution of the derived potential in (15), which yields

$$x_{tr} = \sqrt{(L_1 + L_2)^2 - \frac{K_2(\rho_s^2 - \rho_s\rho_f)}{w\rho_f^2}H_0^2}$$

$$(A22)$$

APPENDIX B

In this Appendix, we describe the derivation of the analytical solution for groundwater level and seawater-freshwater interface in an island, bounded below by a horizontal impermeable bottom, in terms of the potential function instead of water head, and all symbols used are as defined previously.

Solutions for water level and saltwater interface before reclamation
Figure 20a shows the unconfined ground water flow system before reclamation. The potentials in the whole flow region may be obtained by setting C_{ui} equal to 0 in (A1) and (A2), which yields

$$\Phi = \frac{1}{2}K_1 h^2 - \frac{1}{2}K_1 \frac{\rho_s}{\rho_f}H_0^2 \qquad h \geq \frac{\rho_s}{\rho_f}H_0 \qquad (B1)$$

$$\Phi = \frac{1}{2}K_1 \frac{\rho_s}{\rho_s - \rho_f}(h - H_0)^2 \qquad h \leq \frac{\rho_s}{\rho_f}H_0 \qquad (B2)$$

The boundary conditions of the unconfined system before reclamation are

$$h = H_0 \qquad \Phi = 0 \text{ when } x = 0 \text{ or } L_1 \qquad (B3)$$

The governing equation (A7) for the potential throughout the flow region remains valid. The solution is

$$\Phi = \frac{1}{2}w(L_1 x - x^2) \qquad (B4)$$

The heads in the aquifer may be obtained from (B1), (B2), and (B4):

$$h = \sqrt{\frac{w(\rho_s - \rho_f)}{K_1 \rho_s}(L_1 x - x^2)} + H_0 \qquad 0 \le x \le x_{t1} \ or \ x_{t2} \le x \le L_1$$

$$\text{(B5)}$$

$$h = \sqrt{\frac{w}{K_1}(L_1 x - x^2) + \frac{\rho_s}{\rho_f} H_0^2} \qquad x_{t1} \le x \le x_{t2} \qquad \text{(B6)}$$

Set h equal to $\rho_s H_0 / \rho_f$ in either (B1) or (B2) and then substitute this value of the potential in (B4), one has

$$x^2 - L_1 x + \frac{K_1(\rho_s^2 - \rho_s \rho_f)}{w \rho_f^2} H_0^2 = 0 \qquad \text{(B7)}$$

The locations of the tips of saltwater tongues can be readily obtained by solving for x_{t1} and x_{t2} in equation (B7).

Solution for ground water level and saltwater interface after reclamation.
Figure 20b shows the ground water flow system after reclamation. The potentials in the reclaimed portion of the aquifer are defined as

$$\Phi = \frac{1}{2} K_2 h^2 - \frac{1}{2} K_2 \frac{\rho_s}{\rho_f} H_0^2 \qquad h \ge \frac{\rho_s}{\rho_f} H_0 \qquad \text{(B8)}$$

$$\Phi = \frac{1}{2} K_2 \frac{\rho_s}{\rho_s - \rho_f}(h - H_0)^2 \qquad h \le \frac{\rho_s}{\rho_f} H_0 \qquad \text{(B9)}$$

If the hydraulic conductivity jumps at $x = L_1$ then the potential jumps:

$$\Phi^-/\Phi^+ = K_1 / K_2 \qquad \text{(B10)}$$

In order to make the potential continuous throughout the flow region, the potentials in the original aquifer are redefined as

$$\Phi = \frac{1}{2} K_1 h^2 - \frac{1}{2} K_1 \frac{\rho_s}{\rho_f} H_0^2 + \gamma \qquad h \ge \frac{\rho_s}{\rho_f} H_0 \qquad \text{(B11)}$$

$$\Phi = \frac{1}{2} K_1 \frac{\rho_s}{\rho_s - \rho_f} (h - H_0)^2 + \gamma \qquad\qquad h \leq \frac{\rho_s}{\rho_f} H_0 \qquad \text{(B12)}$$

where $\gamma = \Phi^+ - \Phi^- = \dfrac{K_2 - K_1}{K_2} \Phi^+$.

The boundary conditions at $x=0$ and $x=L=L_1+L_2$, are

$$x = 0 \qquad h = H_0 \qquad \Phi = \gamma \qquad\qquad\qquad\qquad \text{(B13)}$$

$$x = L \qquad h = H_0 \qquad \Phi = 0 \qquad\qquad\qquad\qquad \text{(B14)}$$

The governing equation (A7) for the domain remains valid. The solution is

$$\Phi = -\frac{1}{2} w x^2 + (\frac{wL}{2} - \frac{\gamma}{L}) x + \gamma \qquad\qquad\qquad \text{(B15)}$$

The unknown value γ can be obtained from (B15) by computing Φ at $x = L_1$:

$$\gamma = \frac{K_2 - K_1}{K_2} \Phi(L_1) = \frac{K_2 - K_1}{K_2} \left[-\frac{1}{2} w L_1^2 + (\frac{wL}{2} - \frac{\gamma}{L}) L_1 + \gamma \right]$$

The solution is:

$$\gamma = \frac{w L_1 L_2 L (K_2 - K_1)}{2(L_1 K_2 + L_2 K_1)} \qquad\qquad\qquad\qquad \text{(B16)}$$

The heads in the aquifer can be obtained from (B8), (B9), (B11), (B12), and (B15):

$$h = \sqrt{\frac{w(\rho_s - \rho_f)}{K_1 \rho_s} (\beta x - x^2)} + H_0 \qquad\qquad 0 \leq x \leq x_{tr1} \quad \text{(B17)}$$

$$h = \sqrt{\frac{w}{K_1} (\beta x - x^2) + \frac{\rho_s}{\rho_f} H_0^2} \qquad\qquad x_{tr1} \leq x \leq L_1 \quad \text{(B18)}$$

$$h = \sqrt{\frac{w}{K_2} (\beta x - x^2 + 2\gamma / w) + \frac{\rho_s}{\rho_f} H_0^2} \qquad L_1 \leq x \leq x_{tr2} \quad \text{(B19)}$$

$$h = \sqrt{\frac{w(\rho_s - \rho_f)}{K_2 \rho_s}(\beta x - x^2 + 2\gamma/w)} + H_0 \qquad\qquad x_{tr2} \le x \le L \qquad (B20)$$

where $\beta = \left[K_1 L^2 + (K_2 - K_1)L_1^2 \right]/(L_1 K_2 + L_2 K_1)$, and γ is obtained from (B16).

Set h equal to $\rho_s H_0 / \rho_f$ in either (B11) or (B12) and then substitute the derived potential in (B15), one has

$$x^2 - \beta x + \frac{K_1(\rho_s^2 - \rho_s \rho_f)}{w \rho_f^2} H_0^2 = 0 \qquad (B21)$$

The distance from the left coastline to the tip of the left saltwater tongue, x_{tr1}, is one root of equation (B21). Similarly, the location of the tip of the right saltwater tongue, x_{tr2}, can be obtained as a root of the follow equation:

$$x^2 - \beta x + \frac{K_2(\rho_s^2 - \rho_s \rho_f)}{w \rho_f^2} H_0^2 - 2\gamma/w = 0 \qquad (B22)$$

Acknowledgements

The study was supported by the Research Grants Council of the Hong Kong Special Administrative Region, China (HKU 7105/02P).

REFERENCES

Amin, I. E. and Campana, M. E. (1996), A general lumped parameter model for the interpretation of tracer data and transit time calculation in hydrologic systems. *Journal of Hydrology*, **179**(1-4): 1-21.

Barnes RSK (1991), Dilemmas in the theory and practice of biological conservation as exemplified by British coastal lagoons, *Biological Conservation* **55**(3): 315-328.

Bear, J., A.H.-D. Cheng, S. Sorek, D. Ouazar, and I. Herrera, ed. (1999), *Seawater Intrusion in Coastal quifers-Concepts, Methods and Practices*. Dordrecht, The Netherlands: Kluwer Academic Publishers.

Bear, J., and G. Dagan (1964), Some exact solutions of interface problems by means of the hodograph method, *J. Geophys. Res.*, **69**(8): 1563-1572.

Code of Federal Regulations (1988), Title 10--Energy: Department of Energy (Part 960.2). Miamisburg, OH : LEXIS-NEXIS, Division of Reed Elsevier.

Chapuis, R. P., R. Chesnaux. (2006). Travel time to a well pumping an unconfined aquifer without recharge. *Ground Water* **44** (4): 600-603.

Chesnaux, RE, JW Molson, and RP Chapuis (2005), An analytical solution for ground water transit time through unconfined aquifers. *Ground Water*, **43**(4): 511 -517.

Cornaton F., Perrochet P. (2005), Groundwater age, life expectancy and transit time distributions in advective–dispersive systems; 1.Generalized reservoir theory. *Advances in Water Resources*, **29**(9), 1267-1291.

Diersch, H.-J.G.. (2002), FEFLOW finite element subsurface flow and transport simulation system reference manual, WASY Inst. for Water Resour. Plann. and Syst. Res., Berlin.

Ding Guoping (2006), *Impact of deep building foundations on coastal groundwater flow systems*, Ph.D thesis, Univ. of Hong Kong, China.

Etcheverry, D., and P. Perrochet (2000), Direct simulation of groundwater transit-time distribution using the reservoir theory, *Hydrogeol. J.*, **8**: 200-208.

Fetter, C. W. (1972), Position of the saline water interface beneath oceanic islands. *Water Resources Research* **8** (5): 1307-1315.

Fetter, C.W. (1994), *Applied Hydrogeology*. New York: Prentice-Hall.

Fórizs, I., and Deák, J. (1998), Origin of bank filtered groundwater resources covering the drinking water demand of Budapest, Hungary. In: *Application of isotope techniques to investigate groundwater pollution, IAEA-TECDOC-1046*, Vienna, pp.133-165.

Gelhar, L.W., and J.L. Wilson (1974), Ground-water quality modeling. *Ground Water* **12**(6): 399–408.

Glover, R.E. (1964), The pattern of fresh water flow in coastal aquifer. In Sea Water in Coastal Aquifers. *USGS Water-Supply Paper* 1613-C. Reston, Virginia: USGS.

Guo, H.P. and J.J. Jiao (2007), Impact of Coastal Land Reclamation on Ground Water Level and the Sea Water Interface. *Ground Water* 45 (3): 362–367.

Henry, H.R. (1964), Interfaces between salt water and fresh water in coastal aquifers. In Sea Water in Coastal quifers. USGS Water-Supply Paper 1613-C. Reston, Virginia: USGS.

Jiao, J. J. (2000), Modification of regional groundwater regimes by land reclamation. *Hong Kong Geologist* **6**, 29-36.

Jiao, J. J., S. Nandy, and H. Li. (2001), Analytical studies on the impact of reclamation on groundwater flow. *Ground Water* **39**(6): 912-920

Jiao, J.J. (2002), Preliminary conceptual study on impact of land reclamation on groundwater flow and contaminant migration in Penny's Bay, *Hong Kong Geologist,* **8**: 14-20.

Jiao, J. J., and Z. Tang. (1999), An analytical solution of groundwater response to tidal fluctuation in a leaky confined aquifer. *Water Resources Research* **35**(3): 747-751.

Jiao JJ, C Leung, KP Chen, JM Huang, and RQ Huang (2005), Physical and chemical processes in the subsurface system in the land reclaimed from the sea, in *Collections of Coastal Geo-Environment and Urban Development*, P399-407, China Dadi Publishing House, Beijing, China.

Jiao, J. J., XS Wang and S Nandy (2006), Preliminary assessment of the impacts of deep foundations and land reclamation on groundwater flow in a coastal area in Hong Kong, China, *Hydrogeology Journal*, **14** (1-2): 100-114.

Kolditz, O., R. Ratke, H.-J. G. Diershch, and W. Zielke (1998), Coupled Groundwater flow and. transport: 1. Verification of variable density flow and transport models. *Advances in water. resources* **21**: 27–46.

Lee J. Florea and Carol M. Wicks (2001), Solute transport through laboratory-scale karstic aufers. *Journal of Cave and Karst Studies*, **63**(2): 59-66.

Lumb P. (1976), Land reclamation in Hong Kong. In *Proceedings of Residential Workshop on Materials and Methods for Low Coast Road, Rail and Reclamation Works*, Leura, Australia, 299-314.

Mahamood, H. R., and D. R. Twigg (1995), Statistical analysis of water table variations in Bharain. *Quarterly Journal of Engineering Geology* **28**: s63-s74.

Nield, S.P., L. R. Townley, and A. D. Barr (1994), A framework for quantitative analysis of. surface water-groundwater interaction: Flow geometry in vertical section. *Water Resources. Research*, 30(8): 2461-2475.

Ni JR, Borthwick AGL, and Qin HP (2002), Integrated approach to determining postreclamation coastlines, *Journal of Environmental Engineering-Asce*, **128** (6): 543-551.

Noske R. A. (1995), The ecology of mangrove forest birds in peninsular malaysia, *IBIS* 137 (2): 250-263.

Seasholes, N. S. (2003), *Gaining ground: a history of landmaking in Boston*. Cambridge, Mass. : MIT Press.

Simpson, M. J., T. P. Clement, and F. E. Yeomans (2003), Analytical model for computing residence times near a pumping well. *Ground water* **41**(3): 351-354.

Smith, A. J., and J. V. Turner (2001), Density-dependent surface water-groundwater interaction and nutrient discharge in the Swan-Canning Estuary, *Hydrol. Processes*, **15**, 2595-2616.

Smith, A. J. (2004), Mixed convection and density-dependent seawater circulation in coastal aquifers, *Water Resour. Res.*, **40**, W08309, doi:10.1029/2003WR002977.

Strack, O.D.L. (1976), A single-potential solution for regional interface problems in coastal aquifers. *Water Resources Research* **12**(6): 1165-1174.

Strack, O. D. L. (1984), Three-dimensional streamlines in Dupuit-Forchheimer models. *Water Resources Research* **20**(7): 812-822.

Stuyfzand, P. J. (1995), The impact of land reclamation on groundwater quality and future drinking water supply in the Netherlands. *Wat. Sci. Tech.* **31**(8): 47-57.

Suzuki, T. (2003), Economic and geographic backgrounds of land reclamation in Japanese ports. *Marine Pollution Bulletin* **47**:226-229.

Terawaki T, Yoshikawa K, Yoshida G, Uchimura M, and Iseki K (2003), Ecology and restoration techniques for Sargassum beds in the Seto Inland Sea, Japan, *Marine Pollution Bulletin* **47** (1-6): 198-201.

Townley, L.R., and M.R. Davidson (1988), Definition of a capture zone for shallow water table lakes, *Journal of Hydrology*, **104**: 53-76.

Varni, M., and J. Carrera (1998), Simulation of groundwater age distributions,. *Water Resour. Res.*, **34**: 3271–3281.

In: Groundwater Research and Issues   ISBN: 978-1-60456-230-9
Editors: W. B. Porter, C. E. Bennington, pp. 105-129   © 2008 Nova Science Publishers, Inc.

Chapter 4

SENSITIVITY, UNCERTAINTY, AND RELIABILITY IN GROUNDWATER MODELLING

Husam Baalousha[*]

Hawke's Bay Regional Council, Private Bag 6006, Napier, New Zealand

ABSTRACT

Groundwater numerical models are powerful and efficient tools for groundwater management, protection, and remediation. However, groundwater modelling, which requires a huge amount of data, is not an easy endeavor. To build a predictive model, and to get reliable results, input data should be accurate and representative of the real situation in the field. Because of the randomness inherent in nature and the heterogeneity of aquifers, it is very difficult to accurately determine the hydrological properties of the aquifers.

Classical groundwater models usually handle input parameters in a deterministic way, without considering any variability, uncertainty, or randomness in these parameters. Thus, the results of deterministic modelling are questionable.

To account for uncertainty in physical, chemical, and geological data, stochastic modelling is usually used. Many approaches have been developed and used, including sampling approaches, reliability methods, and the Monte Carlo simulation. In this chapter, different approaches of stochastic and probabilistic modelling are introduced and discussed.

1. INTRODUCTION

Groundwater pollution from agricultural activities, industrial and domestic waste has resulted in deterioration of groundwater resources in many regions around the world. These stresses on water resources have increased the concentrations of pollutants in groundwater to high levels. Several processes and chemical reactions occur as the pollutants infiltrate through the

---

[*] E-mail address: Baalousha@web.de, Phone: +64 (0) 6- 8338012, Fax: +64 (0) 6- 8353601

unsaturated zone to the groundwater aquifer, affecting the fate and the concentration of the contaminants in groundwater. To take a successful remediation action, these processes should be understood.

Numerical modelling of groundwater contamination is a powerful tool and can be relied on; however, groundwater modelling is not an easy task. To build a predictive model, and to get reliable results, input data should be accurate and representative of the real system under study. Because of heterogeneity of the aquifers and uncertainties in model input data, including chemical, physical, and hydrogeological parameters, the modelling process turns into a complicated task. In addition, mathematical modelling implies many assumptions and estimations, which increase the uncertainty of the model outputs. To eliminate these problems, input data should be as much and as accurate as possible. Unfortunately, it is very difficult to collect all the required data with a high degree of accuracy. As a result, the output of numerical models always contains a certain degree of uncertainty.

In groundwater modelling, different probabilistic models are used to cope with uncertainty in flow and contaminant transport models (Yen *et al.* 1986; Cawlfield and Sitar 1988; Lioua and Der Yeha 1997).

The most commonly used methods in uncertainty analysis are the sampling methods (e.g. Monte Carlo Simulation, and Latin Hypercube Sampling), the Mean Value Method and Reliability Methods (e.g. First Order Reliability Method).

This chapter outlines these methods and their advantages and disadvantages when used in uncertainty analysis in groundwater studies.

1.1. Uncertainty in Groundwater Modelling

Uncertainty in a parameter reflects the incomplete knowledge of the nature of that parameter. This incompleteness in knowledge (either in information or context) causes model-based predictions to differ from reality in a manner described by some distribution function (Delaurentis and Mavris 2000). Because of the inherent random nature of hydrological parameters in space and time, these parameters have some uncertainties. Modelling of a hydrological system requires estimations and assumptions, and as a result, leads to further uncertainty. There are several sources of uncertainty in any groundwater model, including natural uncertainties, model uncertainties, parameter uncertainties, and data uncertainties (Yen *et. al.* 1986).

1.1.1. Natural Uncertainty

Groundwater systems are inherently stochastic because of the random nature of hydrogeological parameters. This randomness can exist in time or space or in both. The groundwater level time series can be considered as an example of stochastic change over time. The inherent uncertainty in time cannot be eliminated even if the groundwater level-records are good. In this case, an accurate time series of groundwater level cannot be obtained. Uncertainty in hydrological parameters such as hydraulic conductivity is a good example of natural uncertainty in space. In this respect, there is indeed a problem of shortage of measurements. It is too difficult to measure the hydraulic or hydrogeological parameters at every point in the model domain. Moreover, the assumptions, which are usually made in groundwater modelling such as aquifer homogeneity, are not precise. This type of uncertainty is aleatory and cannot be avoided.

1.1.2. Parameter Uncertainty

Parameter uncertainty accompanies input data, which are used in any modelling system. Errors in input parameters lead to this type of uncertainty (Jian and Schilling 1996; Keller *et al.* 2002). Although some input parameters can be measured in the field with some accuracy, the exact value of these parameters cannot be represented at all points.

Errors in measurements can be found as a result of systematic errors in the measuring device. In addition, the estimates that are made to describe the data (i.e. type of distribution, statistical estimation, etc.) increase the uncertainty. Parameter uncertainty is epistemic. As more data is obtained, the parameter uncertainty can be reduced.

1.1.3. Model Uncertainty

By definition, the model is a simplified representation of the real world. The type of model used for this representation, and the different assumptions used to carry out such a model might lead to some errors in the estimated output of the model. There are three types of model uncertainties: conceptual model uncertainty, mathematical model uncertainty, and computer code uncertainty (Zio and Apostolakis 1996). In general, the sources of model uncertainty can be classified as follows:

a) Model Structure Uncertainty
   Model structure uncertainty originates from the assumptions in the model algorithms and model structure. Different models usually produce different solutions as a result of different model structures. The simplifications made in the mathematical computation in the model structure are the main sources of this uncertainty.

b) Model Concept
   The assumptions and simplifications in models, which might be used in the conceptual model, are the sources of conceptual model uncertainty (Hua Lei and Schilling 1996). An example of such simplification in modelling is the assumption that the aquifer is homogenous and isotropic, which is an imperfect assumption and does not exist in reality. Identification of boundary conditions and initial conditions can significantly affect the model result. Errors in boundary and initial conditions increase the model concept uncertainty.

c) Model Resolution
   The model grid size has a great effect on model accuracy. On one hand, if the model grid size is small the accuracy of the model is better (to some extent). On the other hand, as the number of elements increases, the computation time increases, and the round off and truncation errors increase. For this reason, the selection of mesh size is important in terms of accuracy and can result in uncertainty if the mesh is coarse.

2. DETERMINISTIC AND PROBABILISTIC MODELS

Deterministic models are usually used to solve groundwater and contaminant transport problems. These models assume that all the input parameters are known in time and space, and consequently, a deterministic value for each parameter can be assigned. In other words, deterministic models do not consider any uncertainty in input data. This causes errors in the model output because of different sources of uncertainty. Deterministic models might result in poor output, and thus, the objectives of numerical modelling cannot be achieved.

Probabilistic (*or stochastic*) models consider a new formulation of the mathematical equations in a way that produces the output in terms of probability and not a deterministic value. Contrary to deterministic parameters, the uncertain parameters in the probabilistic model are not accurately known, and therefore, they are considered as uncertain parameters or *random variables*. The random variables of the system can be presented in the form of random vector $X = (x_1, x_2, \ldots, x_n)$.

A random variable is a variable whose possible values are numerical outcomes of a random phenomenon. The actual value of a random variable cannot be known. However, the value that the random variable can assume, and the probabilities of these values, are known (Long and Narciso, 1999).

The output of a single trial of a certain random variable cannot be reliably predicted, but the result of a great number of trials can be reliably predicted. The uncertainty characteristics of a random variable can be described by its probability density function (PDF) and/or by its statistics such as first and second moments.

The problem with probabilistic models is that they usually require more computations than deterministic models. The Monte Carlo Simulation (MCS), for example, is one of the probabilistic methods used in groundwater modelling. To get reliable results with MCS, a large number of simulations is required. Some developments of MCS were made to make it more deterministic, and thus, to decrease computations. One of these modifications is the Latin Hypercube Sampling (LHS). The more deterministic the model is, the less the computation required, and the greater uncertainty in model results. It is recommended, therefore, to make a good judgment between computation requirements and the intended accuracy to select the proper modelling method.

3. RELIABILITY THEORY AND LIMIT STATE FUNCTION

The reliability of a system (P_s) is defined as the probability of non-failure in which the resistance of the system (R) exceeds the load (L) (Baalousha 2003). System resistance and load have different meanings according to the problem. In hydrogeology, system failure occurs when the stresses on an aquifer, which can be pollution load or groundwater abstraction, exceed the system resistance. Resistance of the system in this context means the aquifer can be exposed to the stresses without damage or deterioration. If R and L were expressed in stochastic format, the reliability (the probability of non-failure) P_s can be obtained as follows (Figure 1):

$$P_s = P(L_s - R_s) \tag{1}$$

where P is the probability, L_s, and R_s are the load and resistance of the system in stochastic form, respectively. Similarly, the probability of failure P_f is the compliment of the reliability, which can be computed as follows:

$$P_f = P(L_s > R_s) = 1 - P_s \qquad (2)$$

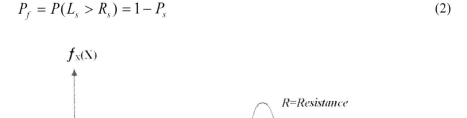

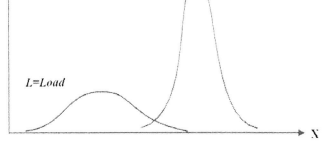

Figure 1. Load-Resistance concept.

Input variables of the hydrogeological system are composed of different parameters. These parameters are classified into two categories: certain and uncertain parameters. For example, groundwater abstraction from a certain well can be precisely measured, so it is considered as a certain parameter. However, the hydraulic conductivity in the aquifer cannot be known because of heterogeneity of the aquifer, and therefore, can be considered as an uncertain parameter or *random variable*. The random variables of the system can be presented in the form of random vector $X=(x_1, x_2, \ldots, x_n)$. The Limit State Function (sometimes called the Performance Function) is a scalar function of the input variables, and it defines the failure domain as shown in Figure 2. In the case of a groundwater flow problem, the limit state function G(X) represents the model output (the value of groundwater head at a certain time and at a certain location). The limit state function is formulated such that $\{x : G(X) = 0\}$, represents the limit state surface. The G-function is expressed with the convention that if $G(x_1, x_2, \ldots, x_n) > 0$, the component survives, whereas if $G(x_1, x_2, \ldots, x_n) \leq 0$, the component fails. Thus, the space of the physical random variables is divided into two domains: the safe, and failure domains. The probability of failure is given by:

$$p_f = p[G(X) \leq 0] = \int_{G(X) \leq 0} f_x(X)dx \qquad (3)$$

where:$f_x(X)$ is the joint probability density function of the random variables $X = (x_1, x_2, \ldots, x_n)$. Because of the difficulties involved in solving the probability function, probabilistic methods were developed to solve Equation (3).

In reliability methods, the objective is to find out the reliability index (β). This index can be used as a measure of comparative reliability, and it is computed as follows:

$$\beta = \left(\frac{\mu_G}{\sigma_G} \right) \qquad (4)$$

where:

μ_G: the mean of the limit state function.
σ_G: the standard deviation of the limit state function.

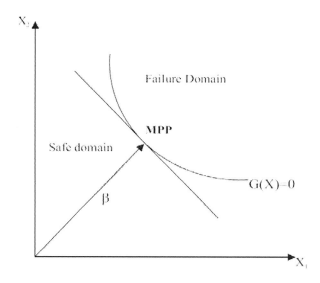

Figure 2. Limit State Function and reliability index.

The reliability (or the probability of non-failure) can be computed as follows:

$$P_s = \Phi(\beta) \qquad (5)$$

where Φ is the standard normal cumulative distribution function.

To get the probability of failure for a given problem by means of Equation (3), the joint probability density function $f_X(X)$ must be constructed, but this is difficult in practice because of the lack of statistical data. Different methods are usually used to solve for the reliability index, as illustrated in the following sections.

3.1. Mean Value Method

The Mean Value method (also called the mean value first order-second moment FOSM) uses the first terms of a Taylor Series Expansion of the performance function to approximate the mean value and the variance of the function (Morgan and Henrion 1990). The method is called second moment because it is the highest order term used in FOSM (Baecher and Christian 2003). The second and higher order terms of the series are truncated. The result of

FOSM is an estimation of the first two statistical moments: the mean and the standard deviation.

Lioua and Der Yeha (1997) have used FOSM to assess the risk of groundwater pollution using a one dimensional groundwater transport model. It has also been used in probability and uncertainty analysis in a geotechnical problem (Graettinger *et al.* 2002).

The Taylor series expansion of the limit state function is:

$$G(X) = G(x_1, x_2, \ldots, x_n)$$

$$G(X) \approx G(X)\big|_{x=\bar{\mu}} + \sum\nolimits_{l=1}^{m} \frac{\partial G(X)}{\partial x_l}\bigg|_{x=\bar{\mu}} (x_l - \mu_l) +$$

$$\frac{1}{2} \sum\nolimits_{j=1}^{m} \sum\nolimits_{l=1}^{m} \frac{\partial^2 G(X)}{\partial x_l \partial x_j}\bigg|_{x=\bar{\mu}} (x_l - \mu_l)(x_j - \mu_j) + H.O.T \qquad (6)$$

where:

μ is the mean of the means of *m* number of random variables; that is, $\mu = \{\mu_1, \mu_2, \ldots, \mu_m\}$, and H.O.T are the higher order terms. Using the expectation value of the variables in Equation (6) and neglecting the higher order terms, the result is:

$$E[G(X)] \approx G(X)_{x=\bar{\mu}} +$$

$$\sum_{i=1}^{m} \frac{\partial G(X)}{\partial x_i}\bigg|_{x=\bar{\mu}} E[(x_i - \mu_i)] + \frac{1}{2!} \sum_{j=1}^{m} \sum_{i=1}^{m} \frac{\partial^2 G(x)}{\partial x_i x_j}\bigg|_{x=\bar{\mu}} E[(x_i - \mu_i)(x_j - \mu_j)] \qquad (7)$$

The variance can be computes as:

$$V[G(X)] = \sigma_g^2 = E\left[(G(X) - E[G(X)])^2\right]$$

$$\approx \sum_{j=1}^{m} \sum_{i=1}^{m} Cov[x_i x_j] \left(\frac{\partial G(X)}{\partial x_i}\bigg|_{x=\bar{\mu}} \right)\left(\frac{\partial G(X)}{\partial x_j}\bigg|_{x=\bar{\mu}} \right) \qquad (8)$$

In case of independent random variables (i.e. no correlation), Equations (7) and (8) can be rewritten as follows:

$$E[G(X)] \approx G(X)\big|_{X=\mu} \qquad (9)$$

$$V[G(X)] = \sigma_g^2 \approx \sum\nolimits_{i=1}^{n} V[x_i] \left(\frac{\partial G(X)}{\partial x_i}\bigg|_{x=\bar{\mu}} \right)^2 \qquad (10)$$

The reliability index can be calculated using Equation (4); that is the expectation value from Equation (7) divided by the standard deviation, which is the square root of the variance from Equation (8).

If the system response, G(X) is normally distributed, the probability of failure can be expressed as follows:

$$P_f = P_r\{G(X) < 0\} = P_r\left\{\frac{G(X) - \mu_G}{\sigma_G} < \frac{0 - \mu_G}{\sigma_G}\right\}$$

$$= \Phi\left(\frac{-\mu_G}{\sigma_G}\right) = \Phi(-\beta) \tag{11}$$

Despite the ease and simplicity in application of the mean value method, it suffers from different drawbacks. FOSM expands the Taylor series around the mean value and not the most probable point. This is the main weakness of the FOSM method (Baecher 2001; Tung, 1999). When applying FOSM in groundwater modelling, the size of the covariance matrices becomes large. Therefore, the uncertainty in parameters must be moderate (Kunstmann *et al.* 2002).

The FOSM method has been shown to be potentially inaccurate for high reliability calculations as well as for highly non-linear performance functions. It assumes assumptions that are not acceptable such as the form of distribution of load and resistance and linear extrapolation (Baecher 2001). If the calculated probability of failure using FOSM is in the extreme tail of a distribution, it is inaccurate. The computed probability by FORM is very sensitive to the formulation of the limit state function.

3.2. The First Order Reliability Method

Hasofer and Lind (Hasofer and Lind 1974) introduced the idea of the First Order Reliability Method (FORM) in the early 70s in structural engineering as alternative to the MCS method.

The FORM method of Hasofer and Lind was further developed by Rackwitz and Fiessler (Rackwitz and Fiessler 1978). It was used to assess the risk of low probability events in structural engineering (Fiessler et. al. 1976). The method was recently used in a hydrological engineering and risk assessment of groundwater pollution (Sitar *et al.* 1987; Hamed and Bedient 1997; Baalousha 2003).

3.2.1. The Hasofer-Lind Approach

To overcome the problem of the Mean Value Method, Hasofer and Lind (1974) proposed another approach, which depends on the expansion about a unique point in the standard space. This method depends on the linear approximation of the performance function at a point on the limit state, called the *design point or the most probable point* MPP. The Hasofer-Lind approach assumes the existence of the critical level of the system performance, which divides the system domain into two parts: the acceptable or safe domain and the unacceptable or failure domain as shown in Figure 3. The probability of system failure can be expressed as:

$$P_f = \int\int ... \int_{G(x)\leq 0} f(x)dx \qquad (12)$$

where G(x)<0 is the system failure domain. Assuming a case of two independent random variables R and L. Let R represent a random variable describing the resistance of the system and L represent a random variable describing the loads on the system. The probability of failure is the probability that R is less than L, and the limit state function is G(R, L): R-L=0

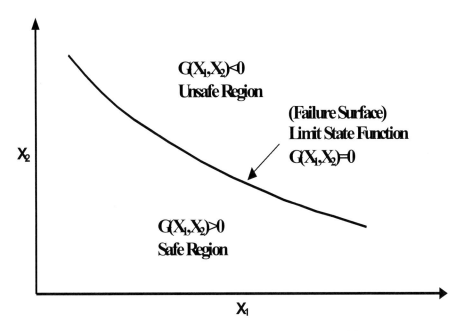

Figure 3. A simple Limit State Function in a physical form.

Hasofer and Lind proposed transformation of the physical parameters of the system variables to eliminate the problem of invariance. The transformation produces a new set of independent random variables. The Hasofer-Lind transformation can be done as follows:

$$u_i = \frac{x_i - \mu_i}{\sigma_i} \qquad (13)$$

where u_i is the transformed variable, μ_i is the mean of a random variable i, and σ_i is the standard deviation of a random variable i.

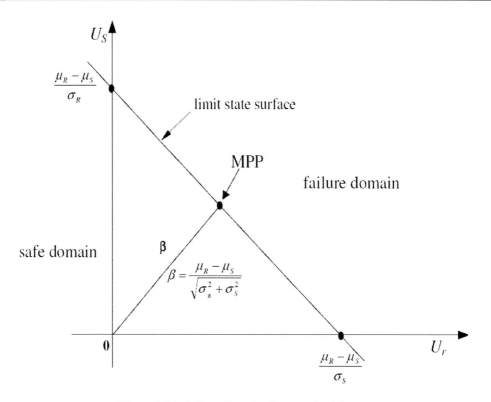

Figure 4. Limit State Function in a standard form.

The new limit state function in the standard form is represented graphically in Figure 4. The minimum distance from the origin to the limit state function is referred to as β.

The point at the limit state function, which is the closest to the origin, is called the most probable point and referred to as MPP. There is a relation between the probability of failure and the safety index as shown in Figure 5. If β increases, the limit state function moves a way from the origin and the probability of failure decreases and vice versa.

From Figure 4, it is clear that the minimum distance from the origin to the limit state function can be calculated as follows:

$$d = \frac{\mu_R - \mu_S}{\sqrt{\sigma_R^2 + \sigma_S^2}} = \frac{\mu_G}{\sigma_G} = \beta \tag{14}$$

Figure 5 shows the limit state function, safety index, and MPP in three-dimensional view. As shown in Figure 4 and 5, the limit state function is independent normally distributed function. Consequently, the probability of failure equals $P_f = \Phi(-\beta)$.

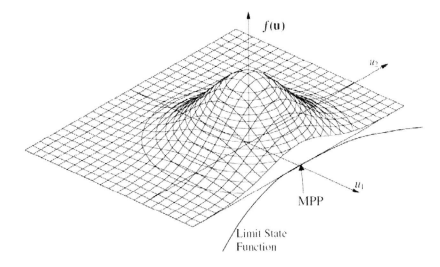

Figure 5. Safety index and MPP point in three-dimensional view.

The difficulty in determination of the design point (or the most probable point) is the determination of the minimum distance from the origin point u* to the limit state function. The general expression of the minimum distance is given by (Shinozuka 1983):

$$\beta = \frac{-\nabla^{*T}u^*}{(\nabla^{*T}\nabla)^{1/2}} = \frac{-\nabla^{*T}u_i^*\left(\dfrac{\partial G}{\partial u_i}\right)\bigg|_{u=u^*}}{\sqrt{\displaystyle\sum_{i=1}^{n}\left(\dfrac{\partial G}{\partial u_i}\right)^2}\bigg|_{u=u^*}} \tag{15}$$

where:

$$\nabla^{*T} = \left(\frac{\partial G}{\partial u_1},\frac{\partial G}{\partial u_2},\cdots\cdots,\frac{\partial G}{\partial u_n}\right)_{u=u^*}$$

The following steps describe a procedure for locating the most probable point based on Hasofer-Lind approach. These steps were presented by Ang and Tang (1984).

The first step is solving FOSM using any arbitrary starting point (usually the mean value):

$$G(X) = G(x_1,x_2,\cdots,x_n)$$
$$\approx G(X)\big|_{x=x^*} + \sum_{i=1}^{n}\frac{\partial G(X)}{\partial x_i}\bigg|_{x=x^*}(x_i - x_i^*) + H.O.T \tag{16}$$

As x* is located on the failure surface then $G(x)_{x=x^*}=0$. Using Hasofer-Lind transformation as presented in Equation (13), and neglecting the high order term, then the last term in the right-hand side of the previous equation ($x_i - x_i^*$) can be written as:

$$x_i - x_i^* = (\sigma_i u_i - \mu_i) - (\sigma_i u_i^* - \mu_i) = \sigma_i (u_i - u_i^*) \tag{17}$$

and the gradient of the limit state function in the standardised form can be expressed using chain rules as follows:

$$\frac{\partial G(X)}{\partial x_i} = \frac{\partial G(u)}{\partial u_i}\left(\frac{\partial u_i}{\partial x_i}\right) = \frac{1}{\sigma_i}\left(\frac{\partial G(u)}{\partial u_i}\right) \tag{18}$$

The mean value at the design point (MPP) is:

$$\mu_G \approx -\sum_{i=1}^n u_i^*\left(\frac{\partial G(u)}{\partial u_i}\right)_{u=u^*} \tag{19}$$

and the variance equals:

$$\sigma_G^2 \approx \sum_{i=1}^n \left(\frac{\partial G(u)}{\partial u_i}\right)_{u=u^*} \tag{20}$$

Finally, the reliability index equals:

$$\beta = \frac{\mu_G}{\sigma_G} \tag{21}$$

where μ_G and σ_G are the mean and the standard deviation at the design point respectively. The probability of failure is:

$$P_f = \Phi\left(\frac{-\mu_G}{\sigma_G}\right) \tag{22}$$

where Φ is the cumulative density function.

3.2.2. The Rackwitz-Fiessler Approach

The main drawback of the Hasofer-Lind approach is that it does not consider the availability of detailed probability density function information. Rackwitz and Fiessler (Rackwitz and Fiessler 1978) proposed one approach improving the accuracy of the reliability index based on the distribution of the variables. The Hasofer-Lind approach assumed that the system response can be accurately represented by a linear combination of Gaussian distributed random variables. However, if the distribution of the random variables is not Gaussian, the approximation can have a significant error. Figure 6 shows the approximation of Weibull probability density function with a normal probability density function. The accuracy is obviously very limited in the tail regions, and there is also no guarantee that the approximation is adequate even in the central regions of the probability density functions. The Rackwitz-Fiessler approach forces the two density functions to have similar statistical properties in the area of primary interest in the probabilistic analysis; specifically in the region of the MPP.

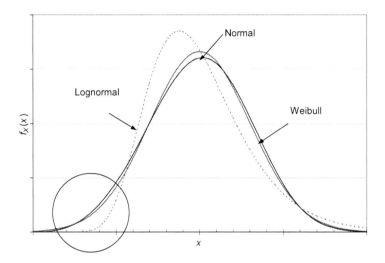

Figure 6. Rackwitz-Fiessler approach.

For any cumulative density function F(X), The Rackwitz-Fiessler approach aimed at finding the mean μ' and standard deviation σ' of an equivalent Gaussian density function such that the cumulative density function and the probability density function at the MPP are both equivalent for the original and the transformed functions (Fiessler *et al.* 1976).

The Rackwitz-Fiessler transformation equations are:

$$F(x*) = \Phi\left(\frac{x * - \mu'}{\sigma'}\right)$$

(23)

$$f(x*) = \frac{1}{\sigma'}\phi\left(\frac{x * - \mu'}{\sigma'}\right)$$

(24)

where:

$\Phi(.)$: the standard normal cumulative distribution function

$\phi(.)$: the standard normal the probability density function

x*: the design point

Solving Equations (23) and (24) for the equivalent mean and standard deviation the result is:

$$\mu' = x^* - \sigma' \Phi^{-1}\left[F(x^*)\right] \qquad (25)$$

$$\sigma' = \frac{\phi\left(\Phi^{-1}\left[F(X^*)\right]\right)}{f(x^*)} \qquad (26)$$

Equations (25) and (26) are used to compute the equivalent first and second moments and then substituted into the procedure of location the most probable point. Then the iterative procedure is used to obtain the design point.

The First Order Reliability Method (FORM)) can be summarised in some steps as illustrated below. As for any FORM optimisation algorithm, the starting point should be first specified. This is usually the mean point.

1) Conversion of all the input variables $X = (x_1, x_2, \ldots, x_n)$ to standard normal variables $U = (u_1, u_2, \ldots, u_n)$ based on Equations 25, and 26.
2) Evaluating the partial derivative at the current design point x^*; that is:

$$\left(\frac{\partial G(u)}{\partial u_i}\right)_{u=u^*}$$

3) Evaluating the performance function and the associated gradient at the current design point.
4) Moving to the new design point according to the following equation:

$$u^*_{i+1} = \frac{1}{(\nabla^*)^T \nabla^*}\left[(\nabla^*)^T u^* - g(u^*)\right]\nabla^* \qquad (27)$$

where ∇ is the gradient vector of variables evaluated at the design point, and T means the transposed matrix.
5) The solution should be revised based on the new value of u
6) If the new value of u is far from the old one, steps from 1 to 3 should be repeated till the convergence achieved.
7) Calculation of the safety index β can be done as follows:

$$\beta = \left[\frac{-\nabla^*}{|\nabla^*|}\right]u^* \qquad (28)$$

3.2.3. Nataf Transformation

The Nataf Transformation (Kiureghian and Liu, 1985) has been widely used in reliability analysis. This approach was developed by Liu and Kiureghian (Liu and Kiureghian, 1986).

Based on that approach, the non-normal correlated random variables can be transformed to the corresponding normal standard variable as follows:

$$Z_i = \Phi^{-1}\left[F_x^i(X_i)\right] \qquad \text{for } i = 1, 2, \ldots, n \tag{29}$$

where:

Z_i: the equivalent standard normal random variable.

$F_x^i(X_i)$: the Cumulative Distribution Function (CDF) of the random variable x_i.

$\Phi^{-1}(.)$: the inverse of the standard normal distribution function.

The normal transform at the design point x, satisfies the following condition:

$$F_i(x_i^*) = \Phi\left(\frac{x_i^* - \mu_{iN}^*}{\sigma_{iN}^*}\right) \tag{30}$$

where:

μ_{iN}: is the equivalent mean value in the normal space at iteration i.

σ_{iN}: is the equivalent standard deviation in the normal space at iteration i.

Based on equations 3.2 and 3.3, the following equation can be derived:

$$\mu_{iN}^* = x_i^* - z_i^* \sigma_{iN}^* \tag{31}$$

where: x_i^* is the random variable at iteration I, z is the standard normal quantile.

The equivalent standard deviation in the normal space can be obtained by the differentiation of both sides of Equation (30). The result is:

$$\sigma_{iN}^* = \frac{\Phi(z_i)}{f_i(x_i^*)} \tag{32}$$

It is clear that the equivalent values of the mean and standard deviation of any variable depend on the expansion point x at that iteration.

For correlated non-normal random variables, Nataf's bivariate distribution model is:

$$\rho_{ij} = \int_{-x}^{x} \int_{-x}^{x} \left(\frac{x_i - \mu_i}{\sigma_i}\right)\left(\frac{x_j - \mu_j}{\sigma_j}\right) \Phi_{ij}(z_i z_j \mid \rho_{ij}^*) dz_i dz_j \tag{33}$$

where:

ρ_{ij} : The correlation coefficient of random variables x_i and x_j in the original space

ρ_{ij} : The correlation coefficient of random variables x_i and x_j in the normal transformed space

Φ_{ij} : The standard cumulative density function.

The only unknown in the above equation is ρ_{ij} . Kiureghian and Liu (Kiureghian and Liu, 1985) have developed a set of empirical formulas to solve the above equation. This formula relates the correlation coefficient in the standard space to that one in the physical space as follows:

$$\rho_{ij}^{*} = T_{ij} * \rho_{ij} \tag{34}$$

where T is a transformation factor that depends on the marginal distribution and correlation of the two random variables under consideration.

As a result, a total of 54 formulas for 10 different distributions were developed and divided into 5 categories. More details about these formulas can be found in (Tung 1999).

3.3. Monte Carlo Simulation

The theoretical foundation of this method had been known for a long time. However, because simulation of random variables by hand is a laborious process, use of the Monte Carlo method as a universal numerical technique became practical only with the advent of computers.

The Monte Carlo simulation has been used as a probabilistic uncertainty propagation technique in different environmental problems, and uncertainty analysis (e.g. Kuczera and Prent 1998; Pet-Armacost e. al. 1999). Bekesi and Mcconchie (1999) have used Monte Carlo to estimate the groundwater recharge in New Zealand. This method was used in the assessment of uncertainty in modelling (Kuczera and Prent 1998), and sensitivity analysis (Pet-Armacost et al., 1999). Many problems, such as optimisation and simulation of fluid movement and sensitivity analysis, are often solved through Monte Carlo simulation (Sobol 1994).

The principle of MCS is simple. Let $f(X)$ denote a vector of random variables with a joint probability density function PDF= $f(x)$. In groundwater models, random variables may be hydraulic conductivity, recharge, etc. Thus, the model output h=g(X) is the groundwater head or the contamination level for example at a certain time and location. In this case, the general form of the multiple-dimensional integral to be solved can be written as:

$$\mu_{h} = E[h] = E[g(X)] = \int_{R} g(X)f(X)dx \tag{35}$$

where R is the probability space R $\in$ [0,1], E is the statistical expectation, and μ_{h} is the mean value. The variance equals:

$$\sigma^{2} = E([g(X) - \mu_{h}]^{2}) = \int_{R} [g(X) - \mu_{h}]^{2} f(X)dx \tag{36}$$

It is very difficult to evaluate the integral in Equation (35) using analytical methods. MCS was employed to solve it. In crude MCS, n independent samples x_1, x_2,..., x_n are generated from the density function $f(x)$ over a uniform space [0,1]. Thus, the MCS estimation of the integral in Equation (35) is:

$$E[f(X)]_{MC} = \mu_{MC} = \frac{1}{n} \sum_{i=1}^{n} g(x_i) \tag{37}$$

According to the strong law of large number, it is known the sum in Equation (37) will converge to the exact value of Equation (35). Also, based on the Limit State Theorem, it is known that:

$$\Pr(\lim_{n \to \infty} \frac{1}{n} \sum_{i=1}^{n} g(x_i)) = \int_R g(x) f(x) \tag{38}$$

from equations (35), and (38) , it is obvious that the mean estimated by MCS is unbiased.
The variance of the estimator is:

$$\text{var}[g(x)]_{MC} = \sigma_{MC}^2 = \frac{1}{n} \sum_{i=1}^{n} (g(x) - \mu_h)^2 \tag{39}$$

Similarly, it has been shown that the variance of the MCS is equal to (Owen, 1998):

$$\sigma_{MC}^2 = \frac{\sigma^2}{n} \tag{40}$$

According to equation (40), the standard error of the estimated integral in equation (35), using MCS is a function of $\frac{1}{\sqrt{n}}$. That means, the error of MCS is inversely proportional to the square root of the number of runs. For instance, to decrease the error by a factor of two, the number of runs should be increased by a factor of four. In other words, to decrease the standard error of MCS, the sampling size should be increased. This can be computationally expensive. A better solution is to employ some technique of variance reduction. These techniques incorporate more information about the analysis directly into the estimator. The methods of variance reduction make the MCS more deterministic, and thus, decrease error.

The advantages of MCS are that it is easy to implement, and requires few data. However, MCS requires a large number of computations to get reliable results. For events with low probability of occurrence, or for contaminant transport problems with a large number of variables, MCS becomes inefficient. The number of samples required to estimate event probability p is in the order of 100/p to get a coefficient of variation of the estimate of 0.10 (Bjerager 1990). The issue is more difficult in the case of complex groundwater problems with too many variables. Therefore, it is not efficient to use MCS in real groundwater problems since they always have a large number of variables.

Figure 7 shows the procedure of generating random variables in MCS. The generated random number is between 0 and 1, and it is uniformly distributed. The generated random number (u_i) should be converted to the equivalent value (x_i) as follows:

$$F_x(x) = u_i \qquad (41)$$

Equation (41)can be solved as follows:

$$x_i = F_x^{-1}(u_i) \qquad (42)$$

Where:
F^{-1}_x is the inverse cumulative distribution function and x is the random variable.

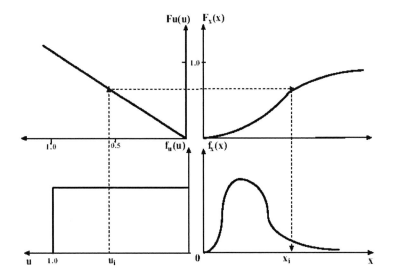

Figure 7. Conversion of generated random number to equivalent variable (Baalousha 2003).

So in a case of a log-normal random variable, the cumulative distribution function is:

$$\phi(x) = \phi\left(\frac{\ln(x_i) - \mu_x}{\sigma_x}\right) \qquad (43)$$

The above equation is equivalent to $\phi^{-1}(u)$; Consequently,

$$\ln(x_i) = \mu_{\ln(x)} + \sigma_{\ln(x)} * \phi^{-1}(u_i) \qquad (44)$$

Thus,

$$x_i = e^{\left[\mu_{\ln(x)} + \sigma_{\ln(x)} * \phi^{-1(u)}\right]} \qquad (45)$$

When using the Monte Carlo method, the problem output should be influenced by random variables. Accordingly, this method is a widely used technique for probabilistic analysis serving two main purposes: validating analytical methods, and solving large, complex systems when analytical approximations are not feasible.

3.4. Latin Hypercube Sampling

Latin Hypercube Sampling (LHS) is one type of stratified MCS. It was first suggested by McKay (McKay *et al.* 1979), and then further developed to improve its efficiency. LHS was used in different computer models for sensitivity and uncertainty analysis (Iman and Conover 1982; Iman and Helton 1988). This method seeks to make the samples more regular than random as in MCS. The idea of the LHS depends on subdivision of the sampling space into N_s number of segment with equal probability and drawing one sample from each segment.

Once the segments are defined, each parameter is then randomised until a value that lies within each probability segment is found. That is, the samples are chosen randomly in such a way that each interval contains one sample. Then, the random numbers for each parameter are combined with the random numbers from the other parameters such that all possible combinations of segments are sampled. The LHS realisations can be generated based on the following formula:

$$ x_{ij} = F^{-1} \left(\frac{\pi_{j(i)} - U_{ij}}{N_s} \right) \tag{46} $$

where:

π_{ij} : the random permutation of $1, \ldots, n$;

n: total number of realisations;
F^{-1}: inverse cumulative probability density function;
U_{ij}: a U[0,1] random variable;
N_s: number of segments.
j:1,2,.... ,k, where k is the dimension of input vector X.

This way, instead of performing a large number of simulations, a fewer number of runs is needed with LHS. In LHS, the sampling points are scattered randomly in each segment. LHS estimator of the function g(x) is:

$$ \mu_{LHS} = \frac{1}{N_s} \sum_{i=1}^{N_s} g(x_i) \tag{47} $$

where:
x_i: the random value selected in interval i.
The relation between variance of LHS and variance of MCS was shown to be equal to (Kollig and Keller 2002):

$$\mathrm{var}[g(x)_{LHS}] \le \frac{n}{n-1}\sigma^2_{MCS} \tag{48}$$

for all $n \ge 2$, where n is the number of intervals. Therefore, LHS is never worse than classical MCS. Stein (1987) has shown that LHS does not reduce the variance relative to simple random sampling, but the reduction depends on the simulated function itself. As a conclusion, LHS can be used with fewer number of runs than MCS and it preserves the accuracy of the results.

3.5. Mean-Value Lattice Sampling

The Mean-Value Lattice Sampling (MVLS) has been developed from LHS and it is used in different stochastic models. The idea of the MVLS depends on subdivision of the probability domain into segments of equal probability in the same way as LHS. The difference from LHS is that MVLS has less randomness. Realisations in MVLS are chosen at the mean value of each segment instead of random value. Thus, the MVLS is more deterministic than LHS. The distribution of all the input random variables should be transformed to standard normal distribution. Then, subdivision of probability space into segments, and determination of the mean value can be done easily. After identification of the mean value of each segment, these points of mean values can be used in sampling

MVLS is more deterministic than LHS, and thus, has less variance in mean estimation. However, this makes MVLS disregard some other areas in the probability domain. The estimator of MVLS is equal to:

$$\mu_{MVLS} = \frac{1}{N_s}\sum_{i=1}^{N_s} g(\mu_{S_i}) \tag{49}$$

where μ_{S_i} is the mean value of segment i. Owen (1992) has shown that the MVLS has less error (i.e. variance) than LHS, and has shown that the variance of Lattice Sampling is not always less than that of LHS. Besides, MVLS does not represent the tail probability very well, and it is not better than LHS if the dimension of the problem is high.

4. SENSITIVITY ANALYSIS

The simplest measure of the sensitivity in reliability analysis is the partial derivative of the reliability index β with respect to the coordination of the design point in the standard space. This value is called α and given by:

$$\alpha = -\frac{\nabla_{u^*}G(u^*)}{\left|\nabla_{u^*}G(u^*)\right|} \tag{50}$$

Based on Equation (50), the vector α gives the rate of change in the reliability index in a standard normal space. There is another expression of alpha sensitivity, which is the same expression as in Equation (50), but in a scaled form (Kiureghian and Ke 1988). This expression of sensitivity vector is called *gamma* and can be expressed as follows:

$$\gamma = \frac{(\nabla_{x^*}\beta)D}{\left|(\nabla_{x^*}\beta)D\right|} = \frac{-\alpha}{\sqrt{D}} \tag{51}$$

where D is the diagonal matrix of standard deviation of the variables. The equation above provides the relative importance of the equal changes of the random variables on the reliability estimate. As a computation check for the above equation, it is noted that:

$$\sum_{i=1}^{n} \gamma^2 = 1, \text{and } -1 \leq \gamma_i \leq 1 \tag{52}$$

The sensitivity of the probability of failure with respect to each random variable can be computed as (Madsen *et al.*, 1986):

$$\frac{\partial P_f}{\partial x_i} = -\alpha_i \phi(\beta) \tag{53}$$

and in the standard form:

$$\frac{\partial P_f}{\partial x_i} = -\frac{\alpha_i \phi(\beta)}{\sigma_i} \tag{54}$$

where $\phi(.)$ is the standard normal probability density function, and P_f is the probability of failure. These sensitivity coefficients reveal the relative importance of each random variable on probability of failure.

As shown above, computation of sensitivity requires evaluating derivatives. The most commonly used method for derivative estimation is the finite difference method. This method, however, suffers from poor accuracy and its result is highly dependent on the step size of the finite difference. In addition, the method is slow and requires solving the function many times to obtain the derivative.

Automatic differentiation is a good alternative for evaluating the gradient vector instead of using the crude finite difference method or manual method. The advantages of automatic differentiation are that it is easy to implement and does not require any knowledge of the original code contents. ADIFOR "Automatic Differentiation of Fortran" (Bischof 1996; Bischof *et. al.* 2002) is a Fortran pre-processor to generate a code that computes the partial derivatives of dependent variables with respect to pre-defined independent variables. There are two approaches for computing derivatives of functions using automatic differentiation: the forward mode and the reverse mode (Bischof *et. al.* 2002). As most groundwater models codes are written in Fortran, ADIFOR is the most appropriate automatic differentiation tool.

The idea of automatic differentiation is dependent on the fact that any computer code, regardless of its length, is composed of set of mathematical operations such as summation, multiplication, etc. Using the chain rule of calculus on all mathematical operations in the code, Adifor obtains the derivative of a dependent variable with respect to the independent one.

Automatic differentiation was used in different groundwater models (Bischof *et. al.* 1994; Baalousha 2003) and the results obtained are accurate up to machine precision.

DISCUSSION

Groundwater models can be classified into deterministic or probabilistic, depending on their way of treating input parameters. Deterministic models are poorly suited when uncertainty in input parameters is high. Probabilistic models account for uncertainty in input parameters, and thus, they are better than deterministic models in this regard. The purpose of this chapter was to document and, when appropriate, identify strengths and weaknesses of methods used for uncertainty analysis in groundwater modelling. The methods of uncertainty analysis can be classified into two main categories: Sampling based methods and reliability methods.

Monte Carlo Simulation (MCS) is the most commonly used sampling-based method. It is straightforward and can be applied with any groundwater problem without restriction. MCS has a major advantage over other methods because it does not restrict the structure of the analysis, so it is used as a base to compare with other methods. As the number of iterations in Monte Carlo grows, the result converges to one value. The result of MCS is unbiased, but a huge number of runs is required to obtain an accurate solution. MCS is a widely used technique for probabilistic analysis serving two main purposes: validating analytical methods, and solving large, complex systems when analytical approximations are not feasible.

Some modifications were done on MCS to make it more deterministic, and thus, to decrease the computation expense. Latin Hypercube Sampling, and Lattice Sampling are modified methods of MCS. Although the computation expense in these methods is less than the crude MCS, but this is in favor of accuracy.

The Mean Value method or the First Order Second Moment method FOSM is a simple method used in probabilistic modelling. The advantage of this method is that it is very easy to implement. However, it has been shown to be potentially inaccurate for high reliability (low probability of failure) calculations (Long and Narciso, 1999). One of the main disadvantages of the FOSM method is that it expands the solution around the mean value. If the limit state function G(X) is non-linear, the result will be inaccurate (Kaymaz *et. al.* 1998).

The First Order Reliability Method FORM is well accepted in the reliability analysis (Kiureghian and Ke 1988; Madsen *et al.* 1986; Tvedt 1988). This method has been applied in different environmental problems and shows good results. The advantage of the FORM method is that it does not requires any pre-conditions to be used, and it can handle both numerical and analytical applications. Moreover, FORM is more computationally efficient than Monte Carlo Simulation in terms of number of iterations. Besides the reliability analysis, FORM method can produce the sensitivity analysis for the result and the importance of the random variables without any extra computations. Different approaches of FORM implementations have been developed depending on the type of transformation. FORM approach requires solving for the gradient, and thus the sensitivity analysis can be done without extra computation. It was found that the First Order Reliability method (FORM) provides a good approximation at much less computational effort compared with the other methods (Baalousha 2003).

Automatic differentiation is a powerful technique for computing derivative codes of groundwater models up to many orders and with precision of the machine code. Automatic differentiation of Fortran (ADIFOR) is good and suits groundwater models as it works with Fortran-written programs. Sensitivity analysis of all input random variables can be very accurately and easily done with automatic differentiation.

REFERENCES

Ang, A., and Tang, W. (1984) *Probability Concepts in Engineering Planning and Design*, Vol. II of Decision, Risk and Reliability, New York: Wiley.

Baalousha, H. (2003) *Risk assessment and uncertainty analysis in groundwater modelling.* PhD. dissertation, Technical University of Aachen (RWTH), Faculty of Civil Engineering. Signatur: Sz8578, Germany

Baecher, G., and Christian, J. (2003) *Reliability and statistics in geotechnical engineering.* England: John Wiley & Sons

Baecher, G. (2001) *Parameters and approximations in geotechnical reliability*, Technical report, University of Maryland.

Bekesi, G., and McConchie, J.A. (1999) Groundwater recharge modelling using the Monte-Carlo technique, Manawatu Region, New Zealand. *Journal of Hydrology*, **224**(3): 137-148.

Bischof, C., Khademi, P., Mauer, A., and Carle, A. (1996) ADIFOR2.0: Automatic differentiating of FORTRAN77 programs. Technical report, *IEEE Computational Science and Engineering* 1996.

Bischof, C., Bücker, M., and Lang, B. (2002) *Automatic differentiation for computational finance computational methods in decision-making, economics and finance.* Kluwer Academic Publishers

Bischof, C., Whiffen, G., Shoemaker, C., Carle, A., and Ross, A. (1994) Application of automatic differentiation to groundwater transport models. *Proceeding of the International Conference on Computational Methods on Water Resources,* Heidelberg

Bjerager, P. (1990) On computation methods for structural reliability analysis. *Structural Safety*, **9**(2): 79–96.

Cawlfield, J., and Sitar, N. (1988) Stochastic finite element analysis of groundwater flow using the first-order reliability method. *IAHS Publication*, **(175):** 191–216.

Yen, B.C., Cheng, .S.T., and Melching, C.S. (1986) *First Order Reliability Analysis. Stochastic And Risk Analysis In Hydraulic Engineering.* Ed by B.C. Yen, Water Resources Publications. Littleton, Co, pp 1-36.

Delaurentis, D.A., and Mavris, D.N. (2000) *Uncertainty modeling and management in multidisciplinary analysis and synthesis.* Technical report, American Institute of

Aeronautics & Astronautics, Aerospace Systems Design Laboratory (ASDL). School of Aerospace Engineering, Georgia Institute of Technology Atlanta, GA 30332-0150.

Fiessler, B., Hawranek, R., and Rackwitz, R. (1976) Numerische Methoden für probabilistische Bemessungsverfahren und Sicherheitsnachweise, Sonderforschungbereich 96, Technische Universität München, [in German].

Graettinger, A.J., Lee, J., and Reeves, H.W. (2002) Efficient conditional modeling for geotechnical uncertainty evaluation. *International Journal For Numerical And Analytical Methods In Geomechanics*, **26**(2): 163–179.

Hamed, M., and Bedient, P. (1997) On the performance of computational methods for the assessment of risk from groundwater contamination. *Groundwater* **35**(4):638–46.

Hasofer, A., and Lind, N. (1974) An exact and invariant first-order reliability format. *Journal of Engineering Mechanics -ASCE* **100**(1):111–21.

Hua Lei, J., and Schilling, W. (1996) Preliminary uncertainty analysis for assessing the predictive uncertainty of hydrologic models. *Water Science and Technology*, **33**(2), 79-90.

Iman, R., and Conover, W. (1982) A distribution-free approach to inducing rank correlation among input variables. *Statistics-Simulation and Computation*, **11**: 311–334.

Iman, R., and Helton, J. (1988) An investigation of uncertainty and sensitivity techniques for computer models. *Risk Analysis*. **8**(1), 71–90.

Jian, H.L., and Schilling, W. (1996) Preliminary uncertainty analysis- a prerequisite for assessing the predictive uncertainty of hydrologic models. *Water Science and Technology*. **33**(2):79–90.

Kaymaz, I., Mcmahon, C., and Meng, X. (1998) Reliability based structural optimisation using the response surface method and Monte Carlo simulation. 8th International Machine Design and Production Conference. Ankara Turkey.

Keller, A., Abbaspour, K.C., and Schulin, R. (2002) Assessment of uncertainty and risk in modeling regional heavy-metal accumulation in agricultural soils. *Journal of. Environmental Quality* **31**:175–187.

Kiureghian, A. Der., and Liu, P.L. (1985) Structural reliability under Incomplete probability information. *Journal of Engineering Mechanics*, **112**(1): 85–104.

Kiureghian, A.Der., and Ke, J.B. (1988): The stochastic finite element method in structural reliability. *Probabilistic Engineering Mechanics*, **3**(2): 83–91.

Kollig, T., and Keller, A. (2002) Efficient multidimensional sampling. Computer Graphics Forum. **21**(3), 557–, doi:10.1111/1467-8659.00706.

Kuczera G., and Prent, E. (1998) Monte Carlo assessment of parameter uncertainty in conceptual catchment models: the metropolis algorithm. *Journal of Hydrology*, **211**(1): 69–85.

Kunstmann, H., Kinzelbach, W., and Siegfried, T. (2002) Conditional first-order second-moment method and its application to the quantification of uncertainty in groundwater modeling. *Water Resources Research*, **38**(4): 6.1-6.15.

Lioua T. and Der Yeha, H. (1997) Conditional expectation for evaluation of risk groundwater flow and solute transport: one dimensional analysis. *Journal of Hydrology*, **199** (3–4): 378–402.

Liu, P.L., and Kiureghian, A.Der. (1986) Multivariate distribution models with Pre-described marginals and covariances. Probabilistic *Engineering Mechanics*, **1**(2): 105–112.

Long, M., and Narciso, J. (1999) Probabilistic design methodology for composite aircraft structures. Technical Report DOT/FAA/AR-99/2, Northrop Grumman Commercial Aircraft Division.

Madsen, H. (1988) Omission sensitivity factors. *Structural Safety*, **5**(1): 35–45.

Madsen, H., Krenk, S., and Lind, N. (1986) *Methods of Structural Safety*. Engle-wood Cliffs, New Jersey: Prentice-Hall, Inc.

McKay, M., Conover, W., and Beckman, R. (1979) A comparison of three methods for selecting values of input variables in the analysis of output from a computer code. *Technometrics*, **21**(2): 239–245.

Morgan MD, and Henrion M. (1990) *Uncertainty: a guide to dealing with uncertainty in quantitative risk and policy analysis.* New York: Cambridge University Press.

Owen, A. (1998) Latin supercube sampling for very high dimensional simulations. *ACM Transactions on Modling and Computer Simulation*, **8**(2), 71–102.

Pet-Armacost, J., Sepulveda, J., and Sakude, M. (1999) Monte Carlo sensitivity analysis of unknown parameters in hazardous materials transportation risk assessment. *Risk Anal;* **19**(6):1173–84.

Rackwitz, R., and Fiessler, B. (1978) Structural reliability under combined random load sequences. *Computers and Structures*, **9**: 489–494

Rajashekhar, M., and Ellingwood, B. (1993) A new look at the response surface approach for structural reliability analysis. *Structural Safety*, **12**(3): 205–225.

Shinozuka, M. (1983) Basic analysis of structural safety. *Journal Of Structural Division, ASCE*, **3**(109): 721-740

Sitar, N., Cawlfield, J., and Kiureghian, A. Der. (1987) First-order reliability approach to stochastic analysis of subsurface flow and contaminant transport. *Water Resources Research*, **23**(5): 794–804.

Sobol, I. (1994) *A primer for the Monte Carlo method*. Crc Press Inc.

Stein, M. (1987) Large sample properties of simulations using Latin Hypercube *Sampling*. *Technometrics*, **29**(2): 143–151.

Tung, Y. (1999) *Risk/reliability-based hydraulic engineering design*. New-York: McGraw-Hill, In: hydraulic design handbook, Chapter 7, pp. 1–56.

Tvedt, L. (1988) Second Order probability by an exact integral, in: *Proceedings Second IFIP Working Conference on Reliability and Optimization of Structural Systems*, pp. 377–384.

Zio, E., and Apostolakis, E. (1996) Two methods for the structured assessment of model uncertainty by experts in performance assessment of radioactive waste repositories. *Reliability Engineering and System Safety*, **54**(2): 225–241.

In: Groundwater Research and Issues   ISBN: 978-1-60456-230-9
Editors: W. B. Porter, C. E. Bennington, pp. 131-145   © 2008 Nova Science Publishers, Inc.

Chapter 5

Environmental Contamination*
Department of Defense Activities Related
to Trichloroethylene, Perchlorate,
and Other Emerging Contaminants

John B. Stephenson
Natural Resources and Environment

What GAO Found

While TCE and perchlorate are both classified by DOD as emerging contaminants, there are important distinctions in how they are regulated and in what is known about their health and environmental effects. Since 1989, EPA has regulated TCE in drinking water. However, health concerns over TCE have been further amplified in recent years after scientific studies have suggested additional risks posed by human exposure to TCE. Unlike TCE, no drinking water standard exists for perchlorate—a fact that has caused much discussion in Congress and elsewhere. Recent Food and Drug Administration data documenting the extent of perchlorate contamination in the nation's food supply has further fueled this debate.

While DOD has clear responsibilities to address TCE because it is subject to EPA's regulatory standard, DOD's responsibilities are less definite for perchlorate due to the lack of such a standard. Nonetheless, perchlorate's designation by DOD as an emerging contaminant has led to some significant control actions. These actions have included responding to requests by EPA and state environmental authorities, which have used a patchwork of statutes, regulations, and general oversight authorities to address perchlorate contamination. Pursuant to its Clean Water Act authorities, for example, Texas required the Navy to reduce perchlorate levels in wastewater discharges at the McGregor Naval Weapons Industrial Reserve Plant to 4 parts per billion (ppb), the lowest level at which perchlorate could be detected at the time. In addition, in the absence of a federal perchlorate standard, at least nine

---

* Excerpted from CRS Report GAO-07-1042T, dated July 12, 2007.

states have established nonregulatory action levels or advisories for perchlorate ranging from 1 ppb to 51 ppb. Nevada, for example, required the Kerr-McGee Chemical site in Henderson to treat groundwater and reduce perchlorate releases to 18 ppb, which is Nevada's action level for perchlorate.

While nonenforceable guidance had existed previously, it was not until EPA adopted its 1989 TCE standard that many DOD facilities began to take concrete action to control the contaminant. According to EPA, for example, 46 sites at Camp Lejeune have since been identified for TCE cleanup. The Navy and EPA have selected remedies for 30 of those sites, and the remaining 16 are under active investigation. Regarding perchlorate, in the absence of a federal standard DOD has implemented its own policies on sampling and cleanup, most recently with its 2006 *Policy on DOD Required Actions Related to Perchlorate*. The policy applies broadly to DOD's active and closed installations and formerly used defense sites within the United States and its territories. It requires testing for perchlorate and certain cleanup actions and directs the department to comply with applicable federal or state promulgated standards, whichever is more stringent. The policy notes, that DOD has established 24 ppb as the current level of concern for managing perchlorate until the promulgation of a formal standard by the states and/or EPA.

Mr. Chairman and Members of the Subcommittee:

We are pleased to be here to discuss our work on the Department of Defense's (DOD) activities associated with emerging contaminants and the cleanup of its hazardous waste sites. DOD defines emerging contaminants as chemicals or materials characterized by (1) a perceived or real threat to human health or environment and (2) a lack of published health standards or a standard that is evolving or being reevaluated. DOD may also classify a contaminant as "emerging" because of the discovery of a new source of contamination, pathway to human exposure, or more-sensitive detection method. Two emerging contaminants—trichloroethylene (TCE) and perchlorate—are of particular concern to DOD because they have significant potential to impact people or DOD's mission.

As we have previously reported,[1] DOD faces the daunting task of cleaning up thousands of military bases and other installations across the country. Many of these sites are contaminated with toxic and radioactive wastes in soil, water, or containers such as underground storage tanks, ordnance and explosives, and unsafe buildings. Identifying and investigating these hazards will take decades, and cleanup will cost many billions of dollars.

In addition to the federal fiscal implications of the large cleanup costs, defense-related contamination problems have economic consequences for individual communities. Many of these formerly used defense sites are now owned by states, local governments, and individuals and used for parks, schools, farms, and homes. Of particular concern are military facilities closed under DOD's Base Realignment and Closure (BRAC) program that are intended to be redeveloped for productive new uses and must generally be cleaned up before conversion. Environmental cleanup is necessary for the transfer of unneeded contaminated property, which becomes available as a result of base closures and realignment.[2] Concerns have risen in recent years within affected communities about the extent to which contamination on these properties could delay or affect the potential for economic redevelopment to replace jobs that were lost as a result of the base closures. While most of the land on bases closed between 1988 and 1995 has been cleaned up and transferred for redevelopment, some has been awaiting cleanup and conversion for many years. Additional bases approved for closure in the 2005 BRAC round will increase the inventory of military properties slated for civilian reuse.

As you requested, my remarks today will focus on (1) the state of knowledge about certain emerging contaminants of concern to the Subcommittee—specifically TCE and perchlorate, (2) DOD's responsibilities for managing emerging contaminants for which federal regulatory standards do not exist, as is the case with perchlorate, and (3) DOD's activities to address the emerging contaminants TCE and perchlorate contamination at its facilities. To address these issues, we relied primarily on our May 2005 report and April 2007 testimony on perchlorate[3] and our May 2007 report and June 2007 testimony on drinking water contamination problems at the Marine Corps Base Camp Lejeune (Camp Lejeune).[4] We also used information from related GAO work on DOD cleanup issues[5] and examined recent data and other information from DOD, the Environmental Protection Agency (EPA), the Food and Drug Administration (FDA), and the states.

In summary, we found the following:

- While TCE and perchlorate are both DOD-classified emerging contaminants, there are important distinctions in the extent to which they are regulated and in what is known about their effects on human health and the environment. TCE, a degreaser for metal parts that DOD has used widely for industrial and maintenance processes, has been found in underground water sources and many surface waters as a result of the manufacture, use, and disposal of the chemical. TCE has been shown to cause headaches and difficulty concentrating at low levels of exposure, whereas high-level exposure may cause dizziness, headaches, nausea, unconsciousness, cancer, and possibly death. As a consequence of these health risks from TCE ingestion, EPA adopted a TCE drinking water standard that became effective in 1989. However, health concerns over TCE have been further amplified in recent years as scientific studies have suggested additional risks posed by human exposure to TCE. In addition, ongoing study of the health affects associated with past exposures on Camp Lejeune may affect DOD's decision whether to settle or deny the pending health claims of former residents. Perchlorate, a primary ingredient in propellant used in the manufacture and firing of rockets and missiles, has been found in drinking water, groundwater, surface water, and soil across the United States. Health studies have shown that it can affect the thyroid gland, which helps to regulate the body's metabolism, and may cause developmental impairments in the fetuses of pregnant women. Unlike TCE, EPA has not set a regulatory standard limiting perchlorate in drinking water—a fact that has caused much discussion in Congress and elsewhere. Recent FDA data documenting extensive, low-level perchlorate contamination in the nation's food supply have further fueled the debate about the extent of perchlorate contamination and its health effects.
- While DOD has certain regulatory compliance responsibilities with regard to emerging contaminants such as TCE that are regulated by EPA or state governments, responsibilities are less definite for other emerging contaminants, such as perchlorate, that lack federal regulatory standards. In the absence of a federal regulatory standard, DOD's designation of perchlorate as an emerging contaminant

indicates its concern about the significant potential impact the chemical has on people or the department's mission. That designation also has resulted in DOD deciding to take certain actions and cleanup efforts even without a federal requirement. While there is no nationwide perchlorate standard, DOD has taken steps to address perchlorate in individual cases in response to EPA regional or state agency actions under various environmental laws such as the Clean Water Act. For example, pursuant to its authority under the Clean Water Act's National Pollutant Discharge Elimination System (NPDES) program, Texas required the Navy to reduce perchlorate levels in wastewater discharges at the McGregor Naval Weapons Industrial Reserve Plant to 4 parts per billion (ppb), the lowest level at which perchlorate could be detected. Also, in the absence of a federal perchlorate standard, at least eight states have established nonregulatory action levels or advisories for perchlorate ranging from 1 ppb to 51 ppb. Nevada, for example, required the Kerr-McGee Chemical site in Henderson to treat groundwater and reduce perchlorate concentration releases to 18 ppb—Nevada's action level.

- DOD is taking a number of actions to address emerging contaminants, including TCE and perchlorate. In 1979, EPA issued nonenforceable guidance establishing "suggested no adverse response levels" for TCE in drinking water. However, the guidance did not suggest actions that public water systems should take if TCE concentrations exceeded those values. Ten years later, EPA's drinking water standard for TCE of 5 ppb became effective. The new standard served as a regulatory basis for many DOD facilities to take concrete actions to control TCE. According to EPA's Region 4 Superfund Director, for example, 46 sites at Camp Lejeune have since been identified for TCE cleanup. The Navy and EPA have selected remedies for 30 of those sites, and the remaining 16 are under active investigation. Regarding perchlorate, in the absence of a federal perchlorate standard, DOD adopted its own policies on sampling and cleanup—specifically a 2003 interim policy followed by a more comprehensive 2006 policy that required more aggressive sampling and, in some cases, cleanup. The 2006 policy applies broadly to DOD's active and closed installations and formerly used defense sites within the United States, its territories and possessions. It directs testing for perchlorate and certain other cleanup actions and directs DOD to comply with applicable federal or state promulgated standards, whichever is more stringent.

THE STATE OF KNOWLEDGE
ABOUT TCE AND PERCHLORATE

While TCE and perchlorate are both DOD-classified emerging contaminants, there are key distinctions between the contaminants that affect the extent to which they are regulated, and the information that may be needed before further steps are taken to protect human health and the environment. Since 1989, a maximum contaminant level (MCL) under the Safe Drinking Water Act has been in place for TCE. In contrast, EPA has not adopted an MCL for perchlorate, although recent government-sponsored studies have raised concerns that even low-levels of exposure to perchlorate may pose serious risks to infants and fetuses of pregnant women.

EPA Has Established a Standard for
TCE and Knowledge Is Evolving

We provided details about EPA's evolving standards for TCE and the evolving knowledge of its health effects in our May 2007 report and June 2007 testimony on issues related to drinking water contamination on Camp Lejeune. TCE is a colorless liquid with a sweet, chloroform-like odor that is used mainly as a degreaser for metal parts. The compound is also a component in adhesives, lubricants, paints, varnishes, paint strippers, and pesticides. At one time, TCE was used as an extraction solvent for cosmetics and drug products and as a dry-cleaning agent; however, its use for these purposes has been discontinued. DOD has used the chemical in a wide variety of industrial and maintenance processes. More recently, the department has used TCE to clean sensitive computer circuit boards in military equipment such as tanks and fixed wing aircraft.

Because TCE is pervasive in the environment, most people are likely to be exposed to TCE by simply eating, drinking, and breathing, according to the Department of Health and Human Services' Agency for Toxic Substances and Disease Registry (ATSDR). Industrial wastewater is the primary source of release of TCE into water systems, but inhalation is the main route of potential environmental exposure to TCE. ATSDR has also reported that TCE has been found in a variety of foods, with the highest levels in meats, at 12 to 16 ppb, and U.S. margarine, at 440 to 3,600 ppb. In fact, HHS's National Health and Nutrition Examination Survey (NHANES) suggested that approximately 10 percent of the population had detectable levels of TCE in their blood.

Inhaling small amounts of TCE may cause headaches, lung irritation, poor coordination, and difficulty concentrating, according ATSDR's Toxicological Profile. Inhaling or drinking liquids containing high levels of TCE may cause nervous system effects, liver and lung damage, abnormal heartbeat, coma, or possibly death. ATSDR also notes that some animal studies suggest that high levels of TCE may cause liver, kidney, or lung cancer, and some studies of people exposed over long periods to high levels of TCE in drinking water or workplace air have shown an increased risk of cancer. ATSDR's Toxicological Profile notes that the National Toxicology Program has determined that TCE is "reasonably anticipated to be a human carcinogen" and the International Agency for Research on Cancer has determined that TCE is probably carcinogenic to humans—specifically, kidney, liver and cervical cancers, Hodgkin's disease, and non-Hodgkin's lymphoma—based on limited evidence of carcinogenicity in humans and additional evidence from studies in experimental animals.

Effective in 1989, EPA adopted an MCL of 5 ppb of TCE in drinking water supplies pursuant to the Safe Drinking Water Act.[6] Despite EPA's regulation of TCE as a drinking water contaminant, concerns over serious long-term effects associated with TCE exposures have prompted additional scrutiny by both governmental and nongovernmental scientific organizations. For example, ATSDR initiated a public health assessment in 1991 to evaluate the possible health risks from exposure to contaminated drinking water on Camp Lejeune. The health concerns over TCE have been further amplified in recent years after scientific studies have suggested additional risks posed by human exposure to TCE. ATSDR is continuing to develop information about the possible long-term health consequences of these potential exposures in a subregistry to the National Exposure Registry specifically for hazardous waste sites.

As we previously reported with respect to Camp Lejeune, those who lived on base likely had a higher risk of inhalation exposure to volatile organic compounds such as TCE, which may be more potent than ingestion exposure. Thus, pregnant women who lived in areas of base housing with contaminated water and conducted activities during which they could inhale water vapor—such as bathing, showering, or washing dishes or clothing—likely faced greater exposure than those who did not live on base but worked on base in areas served by the contaminated drinking water.

Concerns about possible adverse health effects and government actions related to the past drinking water contamination on Camp Lejeune have led to additional activities, including new health studies, claims against the federal government, and federal inquiries. As a consequence of these growing concerns—and of anxiety among affected communities about these health effects and related litigation—ATSDR has undertaken a study to examine whether individuals who were exposed in utero to the contaminated drinking water are more likely to have developed certain childhood cancers or birth defects. This research, once completed later in 2007, is expected to help regulators understand the effects of low levels of TCE in our environment.

In addition, some former residents of Camp Lejeune have filed tort claims and lawsuits against the federal government related to the past drinking water contamination. As of June 2007, about 850 former residents and former employees had filed tort claims with the Department of the Navy related to the past drinking water contamination. According to an official with the U.S. Navy Judge Advocate General—which is handling the claims on behalf of the Department of the Navy—the agency is currently maintaining a database of all claims filed. The official said that the Judge Advocate General is awaiting completion of the latest ATSDR health study before deciding whether to settle or deny the pending claims in order to base its response on as much objective scientific and medical information as possible. According to DOD, any future reassessment of TCE toxicity may result in additional reviews of DOD sites that utilized the former TCE toxicity values, as the action levels for TCE cleanup in the environment may change.

EPA Has Not Established a Standard for Perchlorate

As we discussed in our May 2005 report and April 2007 testimony, EPA has not established a standard for limiting perchlorate concentrations in drinking water under the SDWA. Perchlorate has emerged as a matter of concern because recent studies have shown that it can affect the thyroid gland, which helps to regulate the body's metabolism and may cause developmental impairments in the fetuses of pregnant women. Perchlorate is a primary ingredient in propellant and has been used for decades by the Department of Defense, the National Aeronautics and Space Administration, and the defense industry in manufacturing, testing, and firing missiles and rockets. Other uses include fireworks, fertilizers, and explosives. It is readily dissolved and transported in water and has been found in groundwater, surface water, drinking water, and soil across the country. The sources of perchlorate vary, but the defense and aerospace industries are the greatest known source of contamination.

Scientific information on perchlorate was limited until 1997, when a better detection method became available for perchlorate, and detections (and concern about perchlorate contamination) increased. In 1998, EPA first placed perchlorate on its Contaminant Candidate List, the list of contaminants that are candidates for regulation, but the agency concluded that information was insufficient to determine whether perchlorate should be regulated under the SDWA.[7] EPA listed perchlorate as a priority for further research on health effects and treatment technologies and for collecting occurrence data. In 1999, EPA required water systems to monitor for perchlorate under the Unregulated Contaminant Monitoring Rule to determine the frequency and levels at which it is present in public water supplies nationwide.[8]

Interagency disagreements over the risks of perchlorate exposure led several federal agencies to ask the National Research Council (NRC) of the National Academy of Sciences to evaluate perchlorate's health effects. In 2005, NRC issued a comprehensive review of the health effects of perchlorate ingestion, and it reported that certain levels of exposure may not adversely affect healthy adults. However, the NRC-recommended more studies on the effects of perchlorate exposure in children and pregnant women and recommended a reference dose of 0.0007 milligrams per kilogram per day. In 2005, the EPA adopted the NRC recommended reference dose, which translates to a drinking water equivalent level (DWEL) of 24.5 ppb. If the EPA were to develop a drinking water standard for perchlorate, it would adjust the DWEL to account for other sources of exposure, such as food.

Although EPA has taken some steps to consider a standard, in April 2007 EPA again decided not to regulate perchlorate—citing the need for additional research—and kept perchlorate on its Contaminant Candidate List. Several human studies have shown that thyroid changes occur in human adults at significantly higher concentrations than the amounts typically observed in water supplies. However, more recent studies have since provided new knowledge and raised concerns about potential health risks of low-level exposures, particularly for infants and fetuses. Specifically, in October 2006, researchers from the Centers for Disease Control and Prevention (CDC) published the results of the first large study to examine the relationship between low-level perchlorate exposure and thyroid function in women with lower iodine levels. About 36 percent of U.S. women have these lower iodine levels. The study found decreases in a thyroid hormone that helps regulate the body's metabolism and is needed for proper fetal neural development.

Moreover, in May 2007, FDA released a preliminary exposure assessment because of significant public interest in the issue of perchlorate exposure from food. FDA sampled and tested foods such as tomatoes, carrots, spinach, and cantaloupe; and other high water content foods such as apple and orange juices; vegetables such as cucumbers, green beans, and greens; and seafood such as fish and shrimp for perchlorate and found widespread low-level perchlorate levels in these items. FDA is also planning to publish, in late 2007, an assessment of exposure to perchlorate from foods, based on results from its fiscal year 2005-2006 Total Diet Study—a market basket study that is representative of the U.S. diet.

Some federal funding has been directed to perchlorate studies and cleanup activities. For example, committee reports related to the DOD and EPA appropriations acts of fiscal year 2006 directed some funding for perchlorate cleanup. In the Senate committee report for the Department of Health and Human Services fiscal year 2006 appropriations act, the committee encouraged support for studies on the long-term effects of perchlorate exposure. The Senate committee report for FDA's fiscal year 2006 appropriations act directed FDA to continue conducting surveys of perchlorate in food and bottled water and to report the findings to Congress. In the current Congress, legislation has been introduced that would require EPA to establish a health advisory for perchlorate, as well as requiring public water systems serving more than 10,000 people to test for perchlorate and disclose its presence in annual consumer confidence reports.[9] Other pending legislation would require EPA to establish a national primary drinking water standard for perchlorate.[10]

DOD's Responsibilities to Address Perchlorate and Other Emerging Contaminants Where Federal Regulatory Standards Do Not Exist

DOD has certain responsibilities with regard to emerging contaminants such as TCE that are regulated by EPA or state governments, but its responsibilities and cleanup goals are less definite for emerging contaminants such as perchlorate that lack federal regulatory standards. As we have previously reported, DOD must comply with any cleanup standards and processes under all applicable environmental laws, regulations, and executive orders, including the Comprehensive Environmental Response, Compensation, and Liability Act of 1980 (CERCLA), the Resource Conservation and Recovery Act (RCRA) and the Clean Water Act's National Pollutant Discharge Elimination System (NPDES), and the SDWA. DOD's designation of perchlorate as an emerging contaminant reflects the department's recognition that the chemical has a significant potential impact on people or the Department's mission. DOD's recognition of a substance as an emerging contaminant can lead DOD to decide to take to certain cleanup efforts even in the absence of a federal regulatory standard. In addition, federal laws enacted in fiscal years 2004 and 2005 required DOD to conduct health studies and evaluate perchlorate found at military sites. For example, the Ronald W. Reagan National Defense Authorization Act for fiscal year 2005 stated that the Secretary of Defense should develop a plan for cleaning up perchlorate resulting from DOD activities when the perchlorate poses a health hazard and continue evaluating identified sites.[11]

As we reported in our 2005 perchlorate report, DOD has sometimes responded at the request of EPA and state environmental authorities—which have used a patchwork of statutes, regulations, and general oversight authorities—to act (or require others, including DOD, to act) when perchlorate was deemed to pose a threat to human health and the environment. For example, pursuant to its authority under the Clean Water Act's NPDES program, Texas required the Navy to reduce perchlorate levels in wastewater discharges at the McGregor Naval Weapons Industrial Reserve Plant to 4 parts per billion, the lowest level at which perchlorate could be detected. Similarly, after sampling required as part of a RCRA permit detected perchlorate, Utah officials required ATK Thiokol, an explosives and rocket fuel manufacturer, to install a monitoring well to determine the extent of perchlorate contamination at their facility and take steps to prevent additional releases of perchlorate.

In addition, EPA and state officials also told us during our 2005 review that they have sometimes used their general oversight responsibilities to protect water quality and human health to investigate and sample groundwater and surface water areas for perchlorate. For example, EPA asked Patrick Air Force Base and the Cape Canaveral Air Force Station, Florida, to sample groundwater for perchlorate near rocket launch sites. Previously, both installations had inventoried areas where perchlorate was suspected and conducted limited sampling. DOD officials did not find perchlorate at Patrick Air Force Base and, according to an EPA official, the Department of the Air Force said it would not conduct additional sampling at either installation until there was a federal standard for perchlorate.

Finally, according to EPA, in the absence of a federal perchlorate standard, at least eight states have established nonregulatory action levels or advisories for perchlorate ranging from 1 part per billion to 51 parts per billion. (See table 1.) Massachusetts is the only state to have established a drinking water standard—set at 2 ppb. The California Department of Health Services reports that California will complete the rulemaking for its proposed standard of 6 ppb later this year.[12]

Table 1. States That Have Established Nonregulatory Perchlorate Levels

State	Level (ppb)	Type of Level
Arizona	14	guidance
California	6	notification level
Maryland	1	advisory level
Nevada	18	public notice standard
New Mexico	1	drinking water screening level
Oregon	18	action level
New York	5	drinking water planning level
	18	public notification level
Texas	17	residential protective cleanup level (PCL)
	51	industrial/commercial PCL

Source: EPA and state documents.

States have used these thresholds to identify the level at which some specified action must be taken by DOD and other facilities in their state, in the absence of a federal standard. For example, Oregon initiated in-depth site studies to determine the cause and extent of perchlorate contamination when concentrations of 18 ppb or greater are found. Nevada required the Kerr-McGee Chemical site in Henderson to treat groundwater and reduce perchlorate concentration releases to 18 ppb, which is Nevada's action level for perchlorate. Utah officials told us that while the state did not have a written action level for perchlorate, it may require the responsible party to undertake cleanup activities if perchlorate concentrations exceed 18 ppb.[13]

DOD IS TAKING SEVERAL ACTIONS TO ADDRESS TCE, PERCHLORATE, AND OTHER EMERGING CONTAMINANTS

DOD is undertaking a number of activities to address emerging contaminants in general, including the creation of the Materials of Evolving Regulatory Interest Team (MERIT) to systematically address the health, environmental, and safety concerns associated with emerging contaminants. As noted above, DOD is required to follow EPA regulations for monitoring and cleanup of TCE. In addition, DOD is working with ATSDR, which has projected a December 2007 completion date for its current study of TCE's health effects on pregnant women and their children. In the absence of a federal standard, DOD has adopted its own perchlorate policies for sampling and cleanup activities or is working under applicable state guidelines.

DOD Recently Has Established a Mechanism for Addressing Emerging Contaminants

DOD created MERIT to help address the health, environmental, and safety concerns associated with emerging contaminants. According to DOD, MERIT has focused on materials that have been or are used by DOD, or are under development for use, such as perchlorate, TCE, RDX, DNT and new explosives, naphthalene, perfluorooctanoic acid (PFOA), hexavalent chromium (i.e., chromium VI), beryllium, and nanomaterials. MERIT's initiatives include pollution prevention, detection/analytical methods, human health studies, treatment technologies, lifecycle cost analysis, risk assessment and risk management, and public outreach. Another of MERIT's activities was to create an Emerging Contaminant Action List of materials that DOD has assessed and judged to have a significant potential impact on people or DOD's mission. The current list includes five contaminants—perchlorate, TCE, RDX, naphthalene, and hexavalent chromium. To be placed on the action list, the contaminant will generally have been assessed by MERIT for its impacts on (1) environment, safety, and health (including occupational and public health), (2) cleanup efforts, (3) readiness and training, (4) acquisition, and (5) operation and maintenance activities.

DOD Is Taking Actions to Address TCE

In 1979, EPA issued nonenforceable guidance establishing "suggested no adverse response levels" for TCE in drinking water. These levels provided EPA's estimate of the short- and long-term exposure to TCE in drinking water for which no adverse response would be observed and described the known information about possible health risks for these chemicals. However, the guidance for TCE did not suggest actions that public water systems should take if TCE concentrations exceeded those values. Subsequently, in 1989, EPA set an enforceable MCL for TCE of 5 micrograms per liter, equivalent to 5 ppb in drinking water.

The new standard served as a regulatory basis for many facilities to take concrete action to measure and control TCE. According to EPA's Region 4 Superfund Director, for example, 46 sites on Camp Lejeune have since been identified for TCE cleanup. The Navy and EPA have selected remedies for 30 of those sites, and the remaining 16 are under active investigation. The first Record of Decision was signed in September 1992 and addressed contamination of groundwater in the Hadnot Point Area, one of Camp Lejeune's water systems. Remedies to address groundwater contamination include groundwater "pump and treat" systems, in-situ chemical oxidation, and monitored natural attenuation.[14]

DOD contends that it is aggressively treating TCE as part of its current cleanup program. It notes that the department uses much less TCE than in the past and requires strict handling procedures and pollution prevention measures to prevent exposure to TCE and the release of

TCE into the environment. Specifically, DOD has replaced products containing TCE with other types of cleaning agents such as citrus-based agents, mineral oils and other non-toxic solutions.

DOD Is Sampling for Perchlorate and Taking Cleanup Actions under Certain Conditions

In the absence of a federal perchlorate standard, DOD has adopted its own policies with regard to sampling and cleanup. The 2003 *Interim Policy on Perchlorate Sampling* required the military services—Army, Navy, Air Force, and Marines—to sample on active installations (1) where a reasonable basis existed to suspect that a perchlorate release occurred as a result of DOD activities, and (2) a complete human exposure pathway likely existed or (3) where a particular installation must do so under state laws or applicable federal regulations such as the NPDES permit program. However, DOD's interim policy on perchlorate did not address cleanup responsibilities nor did it address contamination at closed installations.

As we detailed in our previous work, DOD only sampled for perchlorate on closed installations when requested by EPA or a state agency, and only cleaned up active and closed installations when required by a specific environmental law, regulation, or program such as the environmental restoration program at formerly used defense sites. For example, at EPA's request, the U.S. Army Corps of Engineers (Corps) installed monitoring wells and sampled for perchlorate at Camp Bonneville, a closed installation near Vancouver, Washington. Utah state officials also reported to us that DOD removed soil containing perchlorate at the former Wendover Air Force Base in Utah, where the Corps found perchlorate in 2004. However, as we previously reported, DOD cited reluctance to sample on or near active installations because of the lack of a federal regulatory standard for perchlorate.

In the absence of a federal standard, DOD has also worked with individual states on perchlorate sampling and cleanup. For example, in October 2004, DOD and California agreed to prioritize perchlorate sampling at DOD facilities in California, including identifying and prioritizing the investigation of areas on active installations and military sites (1) where the presence of perchlorate is likely based on previous and current defense-related activities and (2) near drinking water sources where perchlorate was found.

In January 2006, DOD updated its policy with the issuance of its *Policy on DOD Required Actions Related to Perchlorate*. The new policy applies broadly to DOD's active and closed installations and formerly used defense sites within the United States, its territories and possessions. It directs DOD to test for perchlorate and take certain cleanup actions. The policy also acknowledges the importance of EPA direction in driving DOD's response to emerging contaminants. It stated, for example, that its adoption of 24 ppb as the current level of concern for managing perchlorate was in response to EPA's adoption of an oral reference dose that translates to a Drinking Water Equivalent Level of 24.5 ppb. The policy also states that when EPA or the states adopt standards for perchlorate, "DOD will comply with applicable state or federal promulgated standards whichever is more stringent."

The 2006 policy directs DOD to test for perchlorate when it is reasonably expected that a release has occurred. If perchlorate levels exceed 24 ppb, a site-specific risk assessment must be conducted. When an assessment indicates that the perchlorate contamination could result in adverse health effects, the site must be prioritized for risk management.[15] DOD uses a relative-risk site evaluation framework across DOD to evaluate the risks posed by one site relative to other sites and to help prioritize environmental restoration work and to allocate resources among sites. The policy also directs DOD's service components to program resources to address perchlorate contamination under four DOD programs—environmental restoration, operational ranges, DOD-owned drinking water systems, and DOD wastewater effluent discharges.

Under the 2006 perchlorate policy, DOD has sampled drinking water, groundwater, and soil where the release of perchlorate may result in human exposure and responded where it has deemed appropriate to protect public health. As we have reported, DOD is responsible for a large number of identified sites with perchlorate contamination, and the department has allotted significant resources to address the problem. According to DOD, sampling for perchlorate has occurred at 258 active DOD installations or facilities. Through fiscal year 2006, DOD reported spending approximately $88 million on perchlorate-related research activities, including $60 million for perchlorate treatment technologies, $9.5 million on health and toxicity studies, and $11.6 million on pollution prevention. Additional funds have been spent on testing technology and cleanup. DOD also claims credit for other efforts, including strict handling procedures to prevent the release of perchlorate into the environment and providing information about perchlorate at DOD facilities and DOD's responses. For example, DOD posts the results of its perchlorate sampling, by state, on MERIT's Web site.[16]

As we have previously reported, DOD must comply with cleanup standards and processes under applicable laws, regulations and executive orders, including EPA drinking water standards and state-level standards. In the absence of a federal perchlorate standard, DOD has also initiated perchlorate response actions to clean up perchlorate contamination at several active and formerly used defense sites under its current perchlorate policy. For example, at Edwards Air Force Base in California, DOD has treated 32 million gallons of ground water under a pilot project for contaminants that include perchlorate. In addition, DOD has removed soil and treated groundwater at the Massachusetts Military Reservation and Camp Bonneville in Washington State.

In conclusion, Mr. Chairman, DOD faces significant challenges, and potentially large costs, in addressing emerging contaminants, particularly in light of the scientific developments and regulatory uncertainties surrounding these chemicals and materials. To help address them, DOD recently identified five emerging contaminants for which it is developing risk management options. As in the case of TCE, DOD took action to address contamination after EPA established an MCL in 1989. DOD has stated that further efforts to address perchlorate would require a regulatory standard from EPA and/or the states. The fact that some states have moved to create such standards complicates the issue for DOD by presenting it with varying cleanup standards across the country.

As the debate over a federal perchlorate standard continues, the recently-issued health studies from CDC and FDA may provide additional weight to the view that the time for such a standard may be approaching. Until one is adopted, DOD will continue to face the challenges of differing regulatory requirements in different states and continuing questions about whether its efforts to control perchlorate contamination are necessary or sufficient to protect human health.

Mr. Chairman, this concludes my prepared statement. I would be happy to respond to any questions that you or Members of the Subcommittee may have at this time.

CONTACTS AND ACKNOWLEDGEMENTS

For further information about this testimony, please contact John Stephenson at (202) 512-3841 or stephensonj@gao.gov. Contact points for our Offices of Congressional Relations and Public Affairs may be found on the last page of this statement. Contributors to this testimony include Steven Elstein, Assistant Director and Terrance Horner, Senior Analyst.

Marc Castellano, Richard Johnson, and Alison O'Neill also made key contributions.

APPENDIX I: SELECTED GAO REPORTS ON DEFENSE-RELATED HAZARDOUS WASTE ISSUES

Defense Health Care: Issues Related To Past Drinking Water Contamination at Marine Corps Base Camp Lejeune, GAO-07-933T (June 12, 2007).

Defense Health Care: Activities Related To Past Drinking Water Contamination at Marine Corps Base Camp Lejeune, GAO-07-276 (May 11, 2007).

Perchlorate: EPA Does Not Systematically Track Incidents of Contamination, GAO-07-797T (April 25, 2007).

Environmental Information: EPA Actions Could Reduce the Availability of Environmental Information to the Public, GAO-07-464T (February 6, 2007).

Military Base Closures: Opportunities Exist to Improve Environmental Cleanup Cost Reporting and to Expedite Transfer of Unneeded Property, GAO-07-166 (January 30, 2007).

Perchlorate: A System to Track Sampling and Cleanup Results Is Needed, GAO-05-462 (May 20, 2005).

Military Base Closures: Updated Status of Prior Base Realignments and Closures, GAO-05-138 (January 13, 2005).

Environmental Contamination: DOD Has Taken Steps to Improve Cleanup Coordination At Former Defense Sites But Clearer Guidance Is Needed to Ensure Consistency, GAO-03-146 (March 28, 2003).

REFERENCES

[1] Appendix I provides a selected bibliography of recent GAO studies on Defense-related hazardous waste issues.

[2] When an installation becomes a BRAC action, the unneeded property is reported as excess. Federal property disposal laws require DOD to first screen excess property for possible reuse by defense and other federal agencies. If no federal agency needs the property, it is declared surplus and is made available to nonfederal parties, including state and local agencies, local redevelopment authorities, and the public.

[3] GAO, *Perchlorate: A System to Track Sampling and Cleanup Results is Needed,* GAO-05-462 (Washington, D.C.: May 20, 2005) and GAO, *Perchlorate: EPA Does Not Systematically Track Incidents of Contamination,* GAO-07-797T (Washington, D.C.: April 25, 2007).

[4] GAO, *Defense Health Care: Activities Related to Past Drinking Water Contamination at Marine Corps Base Camp Lejeune,* GAO-07-276 (Washington, D.C.: May 11, 2007) and GAO, *Issues Related to Past Drinking Water Contamination at Marine Corps Base Camp Lejeune,* GAO-07-933T (Washington, D.C.: June 12, 2007).

[5] GAO, *Military Base Closures: Opportunities Exist to Improve Environmental Cleanup Cost Reporting and to Expedite Transfer of Unneeded Property,* GAO-07-166 (Washington, D.C.: January 30, 2007).

[6] For contaminants that are known or anticipated to occur in public water systems and that the EPA Administrator determines may have an adverse impact on health, the act requires EPA to set a nonenforceable maximum contaminant level goal (MCLG) at which no known or anticipated adverse health effects occur and that allows an adequate margin of safety. Once the MCLG is established, EPA may set an enforceable standard for water as it leaves the treatment plant, the maximum contaminant level (MCL). The MCL generally must be set as close to the MCLG as is feasible using the best technology or other means available, taking costs into consideration.

[7] Under the Safe Drinking Water Act, EPA's determination to regulate a contaminant must be based on findings that: (a) the contaminant may have an adverse effect on the health of persons; (b) the contaminant is known to occur or there is a substantial likelihood that the contaminant will occur in public water systems with a frequency and at levels of public health concern; and (c) in the sole judgment of the Administrator, regulation of such contaminant presents a meaningful opportunity for health risk reduction for persons served by public water systems.

[8] EPA recently determined that it had collected sufficient data and that further monitoring was not needed, 72 *Fed. Reg.* 374, January 4, 2007.

[9] S. 24.

[10] S. 150 and H.R. 1747. A national primary drinking water standard is a legally enforceable standard that applies to public water systems. It sets an MCL or specifies a certain treatment technique for public water systems for a specific contaminant or group of contaminants.

[11] Pub. L. No. 108-375, § 318, 118 Stat. 1811, 1845 (2004).

[12] In September 2006, the California Department of Health Services (CDHS) proposed a primary drinking water standard (in this case a maximum contaminant level, MCL) of 6 ppb for perchlorate. CDHS reports that the completed rulemaking will be submitted to the Office of Administrative Law by August 31, 2007.

[13] According to state and EPA officials, in instances where perchlorate was found, state agencies have sometimes taken steps to minimize human exposure or perform cleanup, or required responsible private parties to do so.

[14] Statement of Franklin Hill, Director of Region 4 Superfund Division, U.S. Environmental Protection Agency, Before the Subcommittee on Oversight and Investigations, Committee on Energy and Commerce, U.S. House of Representatives (June 12, 2007).

[15] DOD's perchlorate website has additional information regarding policy and guidance, http://www.denix.osd.mil/denix/Public/Library/MERIT/Perchlorate/efforts/policy/index .html.

[16] See https://www.denix.osd.mil/denix/Public/Library/MERIT/Perchlorate/index.html.

In: Groundwater Research and Issues   ISBN: 978-1-60456-230-9
Editors: W. B. Porter, C. E. Bennington, pp. 147-177 © 2008 Nova Science Publishers, Inc.

Chapter 6

GROUNDWATER CONTAMINATION: DOD USES AND DEVELOPS A RANGE OF REMEDIATION TECHNOLOGIES TO CLEAN UP MILITARY SITES[*]

U.S. Government Accountability Office

WHAT THE GAO FOUND

DOD has implemented or field-tested all of the 15 types of generally accepted technologies currently available to remediate contaminated groundwater, including several alternatives to pump-and-treat technologies. Some of these technologies, such as bioremediation, introduce nutrients or other materials into the subsurface to stimulate microorganisms in the soil; these microorganisms consume the contaminant or produce byproducts that help break down contaminants into nontoxic or less-hazardous materials. DOD selects the most suitable technology for a given site on the basis of several factors, such as the type of contaminant and location in the subsurface, and the relative cost-effectiveness of a technology for a given site. DOD has identified a number of contaminants of concern at its facilities, each of which varies in its susceptibility to treatment. The table below shows the technologies DOD used to remediate contaminated groundwater.

GAO did not identify any alternative groundwater remediation technologies being used or developed outside DOD that the department has not considered or used. Most of the new approaches developed by commercial vendors and available to DOD generally use novel materials applied to contaminated sites with existing technologies. DOD actively researches and tests new approaches to groundwater remediation largely by developing and promoting the acceptance of innovative remediation technologies. For example, DOD's Strategic Environmental Research and Development Program supports public and private research on contaminants of concern to DOD and innovative methods for their treatment.

---

[*] Excerpted from CRS Report GAO-05-666, dated June 2005.

Technologies DOD Components Used for Groundwater Remediation

Technology	Air Force	Army	Army Corps of Engineers	Defense Logistics Agency	Navy
In-situ					
Air sparging	X	X	X	X	X
Bioremediation	X	X	X	X	X
Enhanced recovery	X			X	X
Chemical treatments	X	X	X	X	X
Monitored natural attenuation	X	X	X	X	X
Multiphase extraction	X	X	X	X	X
Permeable reactive barriers	X	X	X	X	X
Phytoremediation	X	X	X		X
Thermal treatments	X	X	X		X
Ex-situ					
Advanced oxidation processes	X	X	X		X
Air stripping	X	X	X	X	X
Bioreactors		X	X		X
Constructed wetlands	X	X	X		X
Ion exchange	X	X	X		X
Adsorption (mass transfer)	X	X	X	X	X

Source: Department of Defense.

ABBREVIATIONS

CERCLA   Comprehensive Environmental Response, Compensation, and Liability Act
DNAPL   dense nonaqueous phase liquids
DOD   Department of Defense
EPA   Environmental Protection Agency
ESTCP   Environmental Security Technology Certification Program
ITRC   Interstate Technology and Regulatory Council
LNAPL   light nonaqueous phase liquids
RCRA   Resource Conservation and Recovery Act
SERDP   Strategic Environmental Research and Development Program

June 30, 2005
The Honorable John Warner
Chairman
The Honorable Carl Levin
Ranking Minority Member
Committee on Armed Services
United States Senate

The Honorable Duncan L. Hunter
Chairman

The Honorable Ike Skelton

Ranking Minority Member
Committee on Armed Services House of Representatives

The Department of Defense (DOD) has identified close to 6,000 sites at its active, closing, and formerly used defense facilities where the groundwater has been so contaminated by past defense activities and the improper disposal of hazardous wastes that cleanup (remediation) of the site is required.[1roundwater—the water found beneath the earth's surface that fills pores between soil particles, such as sand, clay, and gravel, or that fills cracks in bedrock—accounts for about 50 percent of the nation's municipal, domestic, and agricultural water supply. When groundwater becomes polluted, it can endanger public health or threaten the environment. DOD estimates that cleanup of its contaminated sites will cost billions of dollars and may take decades to complete because of the extent of the contamination and the complexity of groundwater systems.

DOD identifies, investigates, and cleans up contaminated groundwater through its Defense Environmental Restoration Program. This program was established by section 211 of the Superfund Amendments and Reauthorization Act of 1986, which amended the Comprehensive Environmental Response, Compensation, and Liability Act (CERCLA) of 1980. In fiscal year 2004, DOD obligated approximately $1.7 billion for environmental restoration activities, including groundwater remediation, on active, closing, and formerly used defense facilities. Multiple DOD entities—the Air Force, Army, Defense Logistics Agency, and Navy—are responsible for groundwater remediation on active DOD facilities.[2] In addition, the U.S. Army Corps of Engineers (Corps) is responsible for groundwater remediation on properties formerly owned, leased, or used by the military.[3] The Air Force has the greatest number of sites with contaminated groundwater needing remediation, followed by the Navy, Army, Corps, and Defense Logistics Agency.[4] DOD must carry out its groundwater remediation program in a manner consistent with section 120 of CERCLA. Section 120 addresses the cleanup of federal facilities and, among other things, provides for participation in cleanup decisions by the state in which a federal facility is located. Personnel from the installation where the contamination is located work with DOD-hired contractors; regulators (federal, state, local, or tribal); and other stakeholders to evaluate and select appropriate technologies to achieve cleanup goals (e.g., treatment or containment of contaminants). DOD may use a single technology or a combination of technologies to clean up the groundwater at a particular site.

In the past, DOD primarily used traditional "pump-and-treat" technologies to contain or eliminate hazardous contaminants in groundwater. Pump-and-treat technologies extract contaminated groundwater for treatment in above-ground (ex-situ) facilities and are often used to prevent the further spread of contamination in the groundwater. However, according to DOD, the Environmental Protection Agency (EPA), and groundwater remediation experts we consulted, pump-and-treat often is expensive because of long cleanup times, inefficiencies in removing contaminants from the subsurface, and the costs associated with disposing of the contaminant and treated water. Recently, DOD has begun to use alternatives to pump-and-treat technologies that rely on a variety of biological, chemical, or physical processes to treat the contaminated groundwater underground (in-situ).

As directed by Public Law 108-375,[5] and as agreed with your offices, this report (1) describes the groundwater remediation technologies that DOD is currently using or field-testing and (2) examines whether any new groundwater remediation technologies are being used outside the department or are being developed by commercial vendors that may have potential for DOD's use, and the extent to which DOD is researching and developing new approaches to groundwater remediation. In addition, this report provides limited information on the key characteristics, benefits, and limitations of selected groundwater remediation technologies in appendix II.

GAO DEFINITION OF GROUNDWATER REMEDIATION TECHNOLOGY

For this report, we define a technology as a distinct technical method or approach for containing, treating, or removing contaminants found in groundwater.

Any modifications or enhancements to a technology, such as variations in the material or equipment used during treatment, are not considered to be a separate technology.

To determine the range of groundwater remediation technologies DOD is currently using or field-testing, we developed a questionnaire that we sent to the DOD components responsible for DOD's groundwater cleanup efforts—the Air Force, Army, Corps, Defense Logistics Agency, and Navy. In the questionnaire, we listed 15 technologies that are currently available for the treatment of contaminated groundwater and asked the DOD components to indicate which of the technologies they have used and to provide examples of specific groundwater remediation projects.[6] We developed this list of technologies by reviewing existing lists developed by the National Research Council, EPA, and others, as well as by working with a groundwater remediation consulting firm and five nationally recognized groundwater remediation experts. To identify DOD components involved with groundwater remediation activities, we met with department officials responsible for developing policy on groundwater remediation and for researching and developing groundwater remediation technologies. We reviewed documents, reports, and guidance on groundwater remediation from DOD, EPA, and the National Academy of Sciences; and visited an Air Force groundwater remediation project and a facility DOD uses to test innovative groundwater remediation technologies. In addition, we attended a national groundwater remediation conference, and spoke with a number of commercial vendors of groundwater remediation technologies about their products and efforts to develop innovative approaches to groundwater remediation. Information presented in this report is based on publicly available documents and information provided by government officials, independent consultants, and experts. We did not review nonpublic research and development activities that may be ongoing in private laboratories. A more detailed description of our scope and methodology is presented in appendix I. We performed our work from January 2005 through May 2005, in accordance with generally accepted government auditing standards.

RESULTS IN BRIEF

DOD has implemented or field-tested all of the 15 types of generally accepted technologies currently available to remediate groundwater. These various remediation technologies include both in-situ and ex-situ treatments, each of which relies on biological, chemical, or physical processes to clean up groundwater. Of these 15 types of technologies, the Navy reported that it has used all 15 and the Air Force, Army, and Corps have used 14 each. The Defense Logistics Agency, which has significantly fewer sites to clean up than the other DOD components, reported using 9 of the 15 technologies. According to department officials, DOD selects the most suitable technology for a given site on the basis of a number of factors, such as the type of contaminant and its location in the subsurface, and the relative cost-effectiveness of a technology for a given site. DOD has identified a number of contaminants of concern at its facilities, each of which varies in its behavior and susceptibility to treatment by the various technologies. Some of the contaminants, such as chlorinated solvents, can potentially be treated using 14 of the 15 technologies, while others, such as metals, can only be treated effectively with 7 of the 15 technologies. According to analyses conducted by groups such as EPA and the Federal Remediation Technologies Roundtable, the cost-effectiveness and performance of each technology can vary significantly depending, in part, on site-specific conditions. A more detailed description of each of the technologies we identified for cleaning up groundwater is presented in appendix II.

We did not identify any alternative technologies for groundwater remediation being used or developed outside of DOD that it has not considered or employed. However, we did identify a number of new approaches to groundwater remediation being developed by commercial vendors—most of which are also being explored or used by DOD—that are based on modifications of or enhancements to existing technologies. Most of the new approaches involve the use of novel materials applied to contaminated sites using existing technologies. For example, DOD has recently used molasses and vegetable oils at several bioremediation projects to stimulate microorganisms in the subsurface to biodegrade contaminants. Other alternative approaches being developed by commercial vendors usually involve modifying the design of existing technologies. For example, DOD is exploring the use of nanoscale rather than granular sized metals to clean up sites contaminated by chlorinated solvents. In addition, we found that DOD is actively involved in researching and testing new approaches to groundwater remediation, largely through its efforts to develop and promote the acceptance of innovative technologies. For example, DOD maintains several programs—such as the Strategic Environmental Research and Development Program—to support the research, development, and testing of innovative cleanup approaches. This program, a DOD-funded basic and applied research program, supports public and private research on contaminants of concern to DOD and innovative methods for their treatment, as well as a variety of other activities. DOD also pursues innovative solutions to groundwater remediation through its Environmental Security Technology Certification Program. This program field-tests and validates promising innovative environmental technologies and transfers these technologies to the commercial sector. DOD also works with various stakeholders, including the regulatory community, to promote understanding and acceptance of innovative remediation approaches. For example, DOD participates in the Interstate Technology and Regulatory Council, a state-led coalition that works with the private sector, regulators, and other stakeholders to increase the regulatory acceptance of new environmental technologies.

BACKGROUND

DOD sites that require cleanup are often contaminated by many different types of hazardous materials, have contamination in more than one medium (e.g., soil, surface water, or groundwater), and may encompass several acres or even square miles. Groundwater stored in subsurface formations called aquifers can become contaminated in a number of ways. For example, contamination can occur when a liquid hazardous substance soaks down through the soil. Often, groundwater contamination is difficult to address because of the complexity of groundwater systems. The subsurface environment can be composed of numerous layers of diverse types of material—such as sand, gravel, clay, and solid rock—and fractured layers through which groundwater flows. These variations in the subsurface often affect how groundwater flows through a contaminated site and can influence how contaminants are spread and accumulate in the subsurface. Chemical properties of the contaminant also influence its distribution in the subsurface. Typically, contaminated sites consist of a source zone where the bulk of the contaminant is concentrated and a plume of contamination that develops beyond the source of contamination as a result of groundwater flowing through the contaminated site. See figure 1 for an illustration of a site with contaminated groundwater.

Source: Adapted from EPA, Fact Flash #5, Groundwater.

Figure 1. Example of a Site with Contaminated Groundwater.

DOD FACILITIES CAN HAVE SIGNIFICANT GROUNDWATER CONTAMINATION

According to DOD, the Air Force has identified more than 2,500 sites on its active and closing installations with contaminated groundwater; the Navy has identified more than 2,000 sites; the Army has identified about 800 sites; and the Defense Logistics Agency has identified 16 sites. In addition, DOD has identified more than 500 contaminated groundwater sites on formerly used defense sites for which the Corps is responsible for cleanup. Contamination on DOD facilities can pose a threat to military personnel, the public, and the sustainability of DOD's training and testing ranges. DOD first initiated its environmental restoration efforts in 1975. Over the last 10 years, DOD has invested approximately $20 billion for the environmental restoration of contaminated sites, including remediation of contaminated groundwater on and around active, closing, and formerly used defense facilities.[7]

DOD CLEANUP ACTIVITIES GENERALLY FOLLOW THE CERCLA PROCESS

DOD's policies for administering cleanup programs are outlined in its guidance for managing its environmental restoration program and generally follow the CERCLA process for identifying, investigating, and remediating sites contaminated by hazardous materials.[8] According to DOD's guidance, department officials are required to involve EPA, relevant state and local government officials, and the public, among others, at specified points in the cleanup process. See figure 2 for more information on the phases of DOD's environmental cleanup process.

Once DOD identifies potential contamination on one of its facilities, it initiates a preliminary assessment to gather data on the contaminated site. If DOD finds evidence that the site needs remediation, it consults with EPA to determine whether the site qualifies for inclusion on the National Priorities List.[9] If EPA places a DOD facility on the National Priorities List, CERCLA requires DOD to begin the next phase of cleanup within 6 months. During this next phase, called a remedial investigation/feasibility study, DOD characterizes the nature and extent of contamination and evaluates the technical options available for cleaning up the site.

DOD also pursues a remedial investigation/feasibility study for sites that do not qualify for the National Priorities List but require decontamination. Data collected during the remedial investigation influences DOD's development of cleanup goals and evaluation of remediation alternatives. During the feasibility study, often conducted concurrently with the remedial investigation, DOD identifies applicable regulations and determines cleanup standards that will govern its cleanup efforts. CERCLA requires that sites covered by the statute be cleaned up to the extent necessary to protect both human health and the environment. In addition, cleanups must comply with requirements under federal environmental laws that are legally "applicable" or "relevant and appropriate" as well as with state environmental requirements that are more stringent than the federal standards.

Furthermore, CERCLA cleanups must at least attain goals and criteria established under the Safe Drinking Water Act and the Clean Water Act, where such standards are relevant and appropriate under the circumstances.

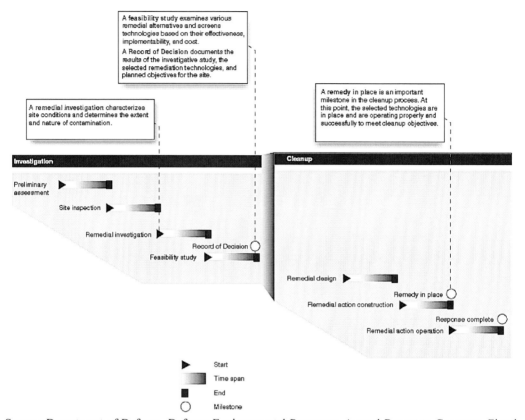

Source: Department of Defense, Defense Environmental Programs, Annual Report to Congress, Fiscal Year 2004.

Note: These phases may overlap or occur simultaneously, but cleanup activities at DOD facilities generally occur in the order shown.

Figure 2. Selected Phases and Milestones in DOD's Environmental Cleanup Process.

Once cleanup standards have been established, DOD considers the merits of various actions to attain cleanup goals. Cleanup actions fall into two broad categories: removal actions and remedial actions. Removal actions are usually short term and are designed to stabilize or clean up a hazardous site that poses an immediate threat to human health or the environment. Remedial actions, which are generally longer term and usually costlier, are aimed at implementing a permanent remedy. Such a remedy may, for example, include the use of groundwater remediation technologies. Also during the feasibility study, DOD identifies and screens various groundwater remediation technologies based on their effectiveness, feasibility, and cost. At the conclusion of the remedial investigation/feasibility study, DOD selects a final plan of action—called a remedial action—and develops a Record of Decision that documents the cleanup objectives, the technologies to be used during cleanup, and the analysis that led to the selection. If EPA and DOD fail to reach mutual agreement on the selection of the remedial action, then EPA selects the remedy. If the cleanup selected leaves any hazardous substances, pollutants, or contaminants at the site, DOD must review the action every 5 years after the initiation of the cleanup.[10] According to DOD policy, this may include determining if an alternative technology or approach is more appropriate than the one in place. DOD continues remediation efforts at a site until the cleanup objectives stated in the Record of Decision are met, a milestone referred to as "response complete." Even if DOD meets the cleanup objectives for a site, in some cases the site may require long-term management and monitoring to ensure that it does not become contaminated from residual sources of pollution.

DOD HAS IMPLEMENTED OR FIELD-TESTED A WIDE RANGE OF TECHNOLOGIES TO REMEDIATE SITES CONTAMINATED WITH GROUNDWATER

DOD has implemented or field-tested all of the 15 types of generally accepted technologies currently available to remediate groundwater. These 15 technologies include 6 ex-situ and 9 in-situ technologies, each of which can be used to treat a variety of contaminants. All of these groundwater remediation technologies rely on a variety of biological, chemical, or physical processes to treat or extract the contaminant. DOD guidance directs department officials to consider cost-effectiveness and performance when selecting technologies for cleanup.

FIFTEEN EX-SITU AND IN-SITU TECHNOLOGIES ARE CURRENTLY AVAILABLE FOR GROUNDWATER CLEANUP

We identified a range of ex-situ and in-situ technologies that DOD can employ to clean up a contaminated groundwater site. Ex-situ technologies rely on a pump-and-treat system to bring the contaminated water above ground so that it can be treated and the contaminants removed. Some ex-situ technologies destroy the contaminant, while others remove the contaminant from the groundwater, which is subsequently disposed of in an approved manner. The decontaminated water can be discharged to surface water, used as part of a public drinking water supply, injected back into the ground, or discharged to a municipal sewage plant. We identified 6 categories of ex-situ technologies:

- *Advanced oxidation processes* often use ultraviolet radiation with oxidizing agents—such as ozone or hydrogen peroxide—to destroy contaminants in water pumped into an above-ground treatment tank.
- *Air stripping* separates volatile contaminants from water by exposing the water to large volumes of air, thus forcing the contaminants to undergo a physical transformation from liquid to vapor (volatilization). There is no destruction of the contaminant; therefore, the contaminant must be removed and disposed of properly.

- *Bioreactors* are above-ground biochemical-processing systems designed to degrade contaminants in water using various microorganisms, an approach similar to that used at a conventional wastewater treatment facility. Contaminated groundwater flows into a tank or basin where it interacts with microorganisms that degrade the contaminant.
- *Constructed wetlands* are artificially built wetland ecosystems that contain organic materials, plants, microbial fauna, and algae that filter or degrade contaminants from the water that is pumped into the wetland.
- *Ion exchange* involves passing contaminated water through a bed of resin media or membrane that exchanges ions in the contaminants, thus neutralizing them into nonhazardous substances.
- *Adsorption (mass transfer)* involves circulating contaminated water through an above-ground treatment vessel containing a sorbent material—such as activated carbon—that removes the contaminant from the water.

Use of Pump-and-Treat Systems

Some groundwater remediation experts believe that pump-and-treat systems may be the best option in situations such as the following:

- the contaminant is located so deep in the subsurface that site characterization and potential remediation can be prohibitively expensive;
- the subsurface is so complex that the effectiveness of in-situ approaches is limited (e.g., highly fractured systems);
- in-situ approaches are not viable or sufficiently proven to remediate a site (e.g., contamination by chlorinated solvents in fractured bedrock);
- the interim cleanup goal is to mitigate risk by containing the contaminant plume (e.g., to protect a public drinking water supply), while an in-situ approach is developed for the site; and
- an ex-situ system is needed to augment or support an in-situ technology.

See figure 3 for an illustration of a pump-and-treat system.
(See app. II for more information on key characteristics of these ex-situ technologies.)

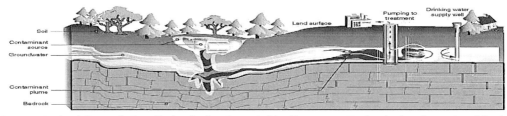

Source: Federal Remediation Technologies Roundtable Treatment Technologies Screening Matrix, 2002.

Figure 3. Example of a Conventional Pump-and-Treat System.

Similarly, we identified nine in-situ technologies that can be used to remediate contaminated groundwater. In contrast to ex-situ technologies, in-situ technologies treat contaminants within the subsurface. Some in-situ technologies—such as bioremediation and chemical treatment—destroy the contaminant within the subsurface by altering the contaminant's chemical structure and converting the toxic chemical to a nontoxic form (e.g., benzene to carbon dioxide). Other in-situ technologies—such as multiphase extraction and enhanced recovery using surfactant flushing—facilitate the removal of the contaminant from the subsurface for treatment above ground. Still other technologies—such as air sparging— combine in-situ treatments with extraction techniques.

- *Air sparging* introduces air or other gases into the subsurface to remove the contamination from the groundwater through volatilization (converting a solid or liquid into a gas or vapor that may be treated at the surface), and in some configurations may also introduce oxygen into the contaminated area to stimulate in-situ biological breakdown (i.e., bioremediation) or ozone to achieve chemical oxidation of the contaminant.
- *Bioremediation* relies on microorganisms living in the subsurface to biologically degrade groundwater contaminants through a process called biodegradation. Bioremediation may be engineered and accomplished in two general ways: (1) stimulating native microorganisms by adding nutrients, oxygen, or other electron acceptors (a process a called biostimulation) or (2) providing supplementary pregrown microorganisms to the contaminated site to augment naturally occurring microorganisms (a process called bioaugmentation).
- *Enhanced recovery using surfactant flushing* involves the injection of active agents known as surfactants[11] into contaminated aquifers to flush the contaminated groundwater toward a pump, which removes the contaminated water and surfactant solution to the surface for treatment and disposal of the contaminants.
- *Chemical treatments* inject various substances into the groundwater that can chemically oxidize or reduce contaminants into less-toxic or nonhazardous materials.
- *Monitored natural attenuation* involves using wells and monitoring equipment in and around a contaminated site to track the natural physical, chemical, and biological degradation of the contaminants. Although not necessarily considered a treatment technology, this approach is often used to monitor contaminant concentrations to ensure that human health and the environment are not threatened.
- *Multiphase extraction* uses a series of pumps and vacuums to simultaneously remove from the subsurface combinations of contaminated groundwater, free product (i.e., liquid contaminants floating on top of groundwater), and hazardous vapors. This technology can be used to remove contaminants from above and below the groundwater table, thereby exposing more of the subsurface for treatment.
- *Permeable reactive barriers* are vertical walls or trenches built into the subsurface that contain a reactive material to intercept and remediate a contaminant plume as the groundwater passes through the barrier.
- *Phytoremediation* relies on the natural hydraulic and metabolic processes of selected vegetation to remove, contain, or reduce the toxicity of environmental contaminants in the groundwater.

- *Thermal treatments* involve either pumping steam into the aquifer or heating groundwater to vaporize or destroy groundwater contaminants. Vaporized contaminants are often removed for treatment using a vacuum extraction system.

(See app. II for more information on key characteristics of these in-situ technologies.)

Although most in-situ technologies have the advantage of treating a contaminant in place, these technologies may afford less certainty about the extent and uniformity of treatment in contaminated areas when compared with some ex-situ technologies. For example, enhanced recovery using surfactant flushing has not been used extensively and has limited data on its remediation effectiveness, whereas air stripping has been widely used for several decades to remove certain contaminants, and its benefits and limitations as a water treatment technology are well-understood. In some cases, a combination of in-situ and ex-situ technologies may be used (either concurrently or successively) to clean up a site if a single technology cannot effectively remediate an entire site with its range of contaminants and subsurface characteristics. According to the National Research Council, integration of technologies is most effective when the weakness of one technology is mitigated by the strength of another technology, thus producing a more efficient and cost-effective solution.[12]

DOD HAS USED THE FULL RANGE OF GROUNDWATER REMEDIATION TECHNOLOGIES IDENTIFIED

As shown in table 1, the DOD components involved in groundwater remediation activities reported using the full range of technologies that we identified as currently available for groundwater remediation. Specifically, the Navy reported that it has used all 15 of the currently available technologies; the Air Force, Army, and Corps reported using 14 each. The Defense Logistics Agency has used 9 of the available technologies for the cleanup of the limited number of contaminated groundwater sites for which it is responsible.

According to department officials, DOD selects the most suitable technology to clean up a contaminated site based on a number of factors, including the type of contaminant, its location and concentration at different levels in the subsurface, and its chemical and physical composition.[13] These officials identified a number of contaminants of concern, such as federally regulated chlorinated solvents (commonly found in metal degreasers) and fuels used for military aircraft and vehicles. DOD officials also consider some other hazardous materials that are not regulated by the federal government—such as the rocket propellant perchlorate— to be contaminants of concern because they are regulated by some states, such as California, where DOD has active, closing, or formerly used defense sites that need groundwater remediation.

Table 1. Technologies DOD Components Used for Groundwater Remediation

Technology	Air Force	Army	Army Corps of Engineers	Defense Logistics Agency	Navy
In-situ					
Air sparginga	X	X	X	X	X
Bioremediationb	X	X	X	X	X
Enhanced recovery/surfactant flushingc	X			X	X
Chemical treatmentsd	X	X	X	X	X
Monitored natural attenuation	X	X			
Multiphase extractione	X	X	X	X	X
Permeable reactive barriersf	X	X	X	X	X
Phytoremediationg	X	X	X		X
Thermal treatmentsh	X	X	X		X
Ex-situ					
Advanced oxidation processesi	X	X	X		X
Air stripping	X	X	X	X	X
Bioreactors		X	X	X	X
Constructed wetlands	X	X	X		X
Ion exchangej	X	X	X		X
Adsorption (mass transfer)	X	X	X	X	X

Source: Department of Defense responses to GAO data collection instrument.

Notes: This table focuses on technologies used to treat contaminants found in groundwater. It excludes technologies used (1) to treat and dispose of the byproducts of groundwater remediation—such as emissions of potentially harmful volatile gases; (2) exclusively to treat contaminated soil (such as soil washing or excavation), although soil remediation is often conducted in conjunction with groundwater remediation; and (3) primarily to physically contain a contaminant—such as soil capping. See appendix II for more information on the key characteristics, benefits, and limitations of each of these technologies.

[a] Includes related remedial approaches and technologies, such as co-metabolic air sparging, oxygen and ozone sparging, in-well air stripping, and soil vapor extraction. Soil vapor extraction, although not technically a groundwater remediation technology, is often used with air sparging to extract or capture emissions that result from treating contaminated groundwater.

[b] Includes related bioremedial approaches, such as bioaugmentation, biostimulation, co-metabolic treatment, enhanced aerobic biodegradation, enhanced anaerobic biodegradation, and biobarriers.

[c] Includes related remedial approaches that use co-solvents to improve the solubility of surfactants in the subsurface, and other technologies, such as hydrofracturing and pneumatic fracturing, that attempt to increase the permeability of the subsurface.

[d] Includes various remedial approaches and technologies that chemically oxidize or reduce contaminants in-situ, as well as the in-situ immobilization and stabilization of soluble metals.

[e] Includes the related technologies of bioslurping and dual-phase extraction.

[f] Includes both biotic and abiotic passive and reactive treatment barriers.

[g] Includes the related technologies of phytostabilization, phytoaccumulation, phytoextraction, rhizofiltration, phytodegradation, rhizosphere degradation, organic pumps, and phytovolatization.

[h] Includes related heating technologies, such as steam flushing, conductive heating, and electrical resistance heating.

[I] Includes the related technologies of ultraviolet oxidation, ultraviolet photolysis, and photocatalysis.

[j] Includes technologies that use ion exchange resins or membranes to remove contaminants from groundwater, including dissolved metals and nitrates.

According to the groundwater remediation experts we consulted, some of DOD's contaminants of concern, such as chlorinated solvents, can potentially be treated using 14 of the 15 technologies, while others, such as metals, can be treated with only 7 of the 15 technologies. For example, many chlorinated solvents do not readily dissolve in water; and because they are often more dense (heavier) than water, they migrate downward and pool at the bottom of aquifers, thereby limiting the number of technologies that can treat them. Alternatively, some contaminants composed of petroleum hydrocarbons (e.g., jet fuel, diesel fuel, and motor gasoline) float on top of the water table because they are less dense (lighter) than water, and technologies such as air sparging or multiphase extraction can often effectively treat or extract them through processes such as volatilization or free product recovery. See table 2 for information on which of the 15 technologies can potentially treat each of DOD's contaminants of concern.

Table 2. Technologies Available for the Treatment of DOD's Contaminants of Concern

Technology	Chlorinated solvents[a]	Explosives[b]	Fuels[c]	Metals[d]	Oxygenates[e]	Propellants[f]
In-situ						
Air sparging	X		X	X	X	X
Bioremediation	X	X	X	X	X	X
Enhanced recovery/surfactant flushing	X		X		X	
Chemical treatments	X	X	X	X	X	X
Monitored natural attenuation	X	X	X	X	X	X
Multiphase extraction	X		X		X	
Permeable reactive barriers	X	X	X	X	X	X
Phytoremediation	X	X	X	X	X	X
Thermal treatments	X		X		X	
Ex-situ						
Advanced oxidation processes	X	X	X	X	X	
Air stripping	X		X		X	
Bioreactors	X	X	X	X	X	X
Constructed wetlands	X	X	X	X	X	X
Ion exchange				X		X
Adsorption (mass transfer)	X	X	X	X	X	

Sources: Department of Defense and several groundwater remediation experts.

Notes: This table presents the contaminants of concern to DOD. Depending on their concentrations, these contaminants can pose health risks to humans. The ability for any one technology to effectively treat a contaminant is greatly influenced by site-specific conditions. Some technologies are generally less effective or currently less utilized to treat contaminants.

[a] Includes, but is not limited to, perchloroethene (PCE), trichloroethene (TCE), dichloroethene (DCE), vinyl chloride (VC), and chloroform (CF). bIncludes, but is not limited to, trinitrotoluene (TNT); dinitrotoluene (DNT); cyclotrimethylene trinitramine, cyclonite, and hexogen (RDX); and octogen and cyclotetramethylene-tetranitramine (HMX). cIncludes gasoline, diesel fuel, jet fuel, and BTEX. BTEX is an acronym for benzene, toluene, ethylbenzene, and xylene—a group of volatile organic compounds commonly found in petroleum hydrocarbons, such as gasoline. dIncludes, but is not limited to, arsenic, barium, cadmium, chromium, copper, lead, mercury, selenium, silver,

and zinc. eIncludes, but is not limited to, oxygen-bearing chemicals that can be added to fuel to bring additional oxygen to the combustion process. These include ethers such as methyl tertiary butyl ether (MTBE) and its related compounds. fIncludes, but is not limited to, materials such as ammonium perchlorate and potassium perchlorate that are used in the manufacturing and testing of solid rocket propellants and other munitions such as flares.

Technology Selection Is Also Influenced by Cost and Performance

According to DOD guidance on groundwater remediation, department officials should consider cost-effectiveness and performance of various groundwater remediation options when selecting the most suitable cleanup technology. A number of factors influence total cleanup costs for a given site, such as how long the cleanup is expected to take and the horizontal and vertical extent of the contamination. In addition, according to the National Research Council, actual cleanup costs associated with each technology depend on site-specific hydrogeologic, geochemical, and contaminant conditions.[14] Thus, a particular technology may be the most cost-effective solution for one site and not necessarily for another similarly contaminated site. The National Research Council and others have also found that performance of most technologies, including time for total cleanup, also depends on complexities within the site's subsurface (i.e., site heterogeneities) as well as contaminant characteristics. For example, the effectiveness of certain in-situ technologies—such as air sparging—decrease as site heterogeneity increases because the air will naturally follow certain pathways that may bypass the contaminant. Similarly, the effectiveness of many in-situ technologies may be limited by the presence of some chlorinated solvents that, if heavier than water, can migrate into inaccessible zones in the subsurface. Alternatively, in-situ thermal treatments that use conductors to heat the soil are not as sensitive to heterogeneity in the subsurface and contaminant characteristics because thermal conductivity varies little with the properties of subsurface materials and certain contaminants are more easily volatilized at elevated temperatures. However, equipment and energy costs may make this approach more costly than other in-situ technologies.

While overall conclusions on the cost-effectiveness of each groundwater remediation technology are difficult to reach, a few groups have attempted to estimate costs for various technologies. For example, EPA has developed a technology cost compendium for several technologies based on cost data from various public and private remediation projects.[15] Similarly, the Federal Remediation Technologies Roundtable—a federal consortium of representatives from DOD, EPA, and other federal agencies—has attempted to evaluate the relative overall cost and performance of selected remediation technologies in general terms.[16] However, according to DOD officials and other experts we consulted, these efforts to compare technologies are of only limited utility because of the site-specific nature of technology decisions.

DOD IS PROACTIVELY USING AND DEVELOPING NEW APPROACHES TO GROUNDWATER REMEDIATION

We did not identify any alternative groundwater remediation technologies being used outside the department that DOD has not already either employed or tested on some scale (laboratory or pilot). However, we did identify a number of new approaches to groundwater remediation being developed by commercial vendors, but these approaches are based on modifications of or enhancements to existing technologies. Most of these new approaches are being used or field-tested by DOD and involve novel materials that are applied to contaminated sites using existing technologies. In addition, we found that DOD is generally aware of new approaches to groundwater remediation, in part through its efforts to develop remediation technologies with the commercial sector. DOD also works with various stakeholders, including the regulatory community, to promote understanding and acceptance of innovative remediation approaches. Some DOD officials and groundwater remediation experts believe additional resources may be needed in order to develop and advance DOD's process for selecting the most appropriate technology at a site.

Most New Approaches Employ Novel Materials or Modifications to Existing Technologies

Most of the new remediation approaches commercial vendors have developed and made available to DOD use existing technologies to apply novel materials to contaminated sites. These materials typically accelerate the breakdown of contaminants through biological or chemical processes. In particular, multiple commercial vendors have developed proprietary compounds used during bioremediation to stimulate microorganisms in the subsurface to biodegrade contaminants. Some of these compounds are designed to slowly release oxygen or other nutrients into the subsurface in an effort to prolong their availability, which microorganisms need to biodegrade the contaminants. DOD has also field-tested several novel compounds for bioremediation that are derived from food-grade materials such as molasses or vegetable oils. These compounds can be injected into the contaminated site using pre-existing wells or other existing techniques such as direct push injection:

- The Army used a compound developed by a commercial vendor to stimulate the bioremediation of chlorinated solvents at a contaminated site at its Rocky Mountain Arsenal. This compound reacted with the contaminated groundwater to produce lactic acid, which native microorganisms used to produce the hydrogen that ultimately led to the biological degradation of the contaminants. In addition, the Air Force reported using oxygen-releasing compounds to stimulate aerobic biodegradation at several of its cleanup sites, including a site in Florida contaminated by spilled fuel.
- DOD has also field-tested the use of molasses during bioremediation to treat chlorinated solvents at Vandenberg and Hanscom Air Force bases. In addition, DOD reported using vegetable oils to stimulate microorganisms in order to treat groundwater contaminated by chlorinated solvents and perchlorate at a variety of

locations, including naval facilities in Massachusetts, Rhode Island, and South Carolina.

Commercial vendors have also developed innovative approaches for chemically treating contaminants in the subsurface. For example, several vendors have developed proprietary approaches for delivering oxidants, such as molecular oxygen and ozone with or without hydrogen peroxide, into the subsurface to achieve in-situ chemical oxidation of a variety of contaminants, including fuels and chlorinated solvents. These oxidants are often delivered underground using variations of existing air sparging technologies and a variety of injection technologies. In addition to achieving in-situ chemical oxidation of target contaminants, the use of ozone with or without hydrogen peroxide can enhance the aerobic biodegradation of contaminants because it increases oxygen levels in the subsurface. Commercial vendors have also developed approaches to directly injecting other chemicals that are oxidizing agents, such as persulfate and permanganate, into the subsurface using existing technologies such as injection wells and direct push-probe technologies.

DOD is exploring with the commercial sector other innovative approaches to groundwater remediation that involve modifying the engineering, design, or application of existing technologies. For example, DOD is currently working with the commercial sector to explore innovative uses of nanoscale metallic materials—such as zero-valent iron and palladium impregnated iron—to improve the efficacy of in-situ chemical treatments of chlorinated solvents commonly found on DOD facilities.[17] In the past, DOD used metallic materials, such as zero-valent iron in granular form, to fill trenches dug into the ground (a form of a permeable reactive barrier) to chemically reduce chlorinated solvent plumes. The iron reacts with chlorinated solvents, transforming them into benign products, such as ethane and ethene. Treating contaminant plumes located deep within the subsurface is often difficult, costly, and technically impossible using this approach. Because of their size, nanoscale particles can be mixed with other materials—such as vegetable oil and water—and injected deep into the subsurface using existing technologies to treat contaminant sources or plumes. Furthermore, nanoscale particles have high surface areas relative to their volume (i.e., more metal is available to contact and react with the contaminants), which will lead to increased rates of reaction and more effective treatment.

DOD SUPPORTS THE DEVELOPMENT OF NEW TECHNOLOGIES WITH THE COMMERCIAL SECTOR THROUGH SEVERAL PROGRAMS

We found that DOD is actively involved in researching and testing new approaches to groundwater remediation, largely through its efforts to develop and promote the acceptance of innovative groundwater remediation technologies. According to the National Research Council, research on innovative remediation technologies is sponsored almost exclusively by federal agencies such as DOD and, in some circumstances, by individual companies and industry groups that have joined with federal agencies in seeking more cost-effective solutions to common problems.[18] In particular, the DOD-funded Strategic Environmental Research and Development Program (SERDP) supports public and private research on contaminants of concern to DOD and innovative methods for their treatment, among other activities. Created in 1990, the program primarily focuses on issues of concern to DOD, although it is jointly managed by DOD, EPA, and the Department of Energy.[19] In fiscal year 2004, SERDP spent about $49 million to fund and manage projects in a variety of areas, including 27 projects related to groundwater remediation.

In response to technology needs and requirements generated by each of the DOD components, SERDP funds research projects in private, public, and academic settings on the fundamentals of contaminant behavior, environmental toxicity, and the advanced development of cost-effective innovative groundwater remediation technologies, among other things. For example, SERDP has funded research projects to examine such issues as the innovative use of vegetable oils for bioremediation; zero-valent iron based bioremediation of explosives; and the behavior of, and treatment options for, several emerging groundwater contaminants not yet regulated by the federal government, such as 1,4-Dioxane (found in solvents), N-Nitrosodimethylamine (found in rocket fuel), and trichloropropane (used as a degreaser and paint stripper). In addition, SERDP holds workshops with the scientific, engineering, academic, regulatory, and DOD-user communities to discuss DOD's issues and identify needs for future research, development, and testing of groundwater remediation techniques.

National Environmental Technology Test Site at Dover Air Force Base.

Source: Dover National Environmental Technology Test Site, Tim McHale.

At Dover Air Force Base, DOD has constructed three double-walled underground test areas (referred to as cells) that enable researchers to inject common soil and groundwater pollutants into a natural geologic setting as test constituents, without allowing the test constituents to come into contact with the surrounding environment. These test cells, known as the Groundwater Remediation Field Laboratory, include one large test cell and several smaller ones, all sharing the same outer containment cell area. The cells are constructed of interlocking steel sheet pilings with sealed grouted joints that extend from the ground's surface to a depth of 40 feet. This safe testing area is in an area with "ideal geology," according to the site program manager, because it has a shallow aquifer contained by a clay layer, which prevents the migration of contaminants. This laboratory is the only place in the United States that offers such a test setting. A variety of technologies have been tested here for cleaning up a range of contaminants. For example, tests for cleanup of TCE are under way using a combination of three technologies: soil vapor extraction, bioremediation, and air stripping.

DOD also pursues innovative solutions to groundwater remediation through its Environmental Security Technology Certification Program (ESTCP). This program, founded in 1995, field-tests and validates promising innovative environmental technologies that attempt to address DOD's highest-priority environmental requirements, including groundwater remediation.[20] Using a process similar to that of SERDP, ESTCP solicits proposals from public and private researchers to field-test laboratory-proven remediation technologies that have broad DOD and market application. Once ESTCP accepts a proposal, it identifies a military partner, which provides a site on a DOD installation where the researcher can fieldtest the technology and document the technology's cost, performance, and reliability. In fiscal year 2004, ESTCP spent about $35 million to fund and manage its program, including 36 projects on groundwater remediation. These projects include the demonstration of an enhanced recovery technology using innovative surfactants, emulsified zero-valent nanoscale iron to treat chlorinated solvents, and an ion exchange technology for the removal and destruction of perchlorate. ESTCP and SERDP have co-located offices and, according to DOD officials, the two programs work together to pursue the development of innovative groundwater remediation technologies from basic research through advanced field-testing and validation. ESTCP often funds the demonstration of technologies that were developed by private or public researchers with financial support from SERDP.

In addition to funding the development of innovative technologies, DOD works with various stakeholders, including the regulatory community, to promote the understanding and acceptance of these technologies. For example, DOD participates in the Interstate Technology and Regulatory Council (ITRC), a state-led coalition that works with the private sector, regulators, and other stakeholders to increase the regulatory acceptance of new environmental technologies. ITRC develops guidance on innovative environmental technologies and sponsors training for regulators and others on technical and regulatory issues related to environmental cleanup technologies and innovative groundwater remediation approaches. According to ITRC, these efforts are designed to help regulators streamline their review process and enable wider acceptance of innovative environmental technologies across state boundaries. In 2004, ITRC and DOD signed a memorandum of understanding on the relationship between the two organizations. As a result of the agreement, DOD now provides several liaisons to the ITRC's board of advisers and helps the group develop materials and training courses on innovative groundwater remediation technologies. According to a DOD official, the department's partnership with ITRC has led to enhanced cooperation among state regulators, DOD personnel, and community stakeholders and increased the deployment of innovative technologies at DOD cleanup sites.

Although DOD is actively involved in the research and development of innovative technologies, some groundwater remediation experts and some DOD officials with whom we consulted believe that additional resources may be needed to develop and advance DOD's process for selecting the most appropriate technology at a site. These individuals believe that a better understanding of the nature and extent of contamination at a site is critical for selecting appropriate technologies for cleanup. Furthermore, these experts and some DOD officials believe that additional resources may be appropriate for examining and improving methods and engineering approaches for optimizing the performance of the 15 types of groundwater remediation technologies that are currently available. Other groundwater remediation experts and some DOD officials suggested that more resources may be needed to further develop innovative approaches to emerging groundwater remediation issues, and to educate DOD personnel and regulators on these approaches.

AGENCY COMMENTS

DOD generally agreed with the content of the report, stating that the report is an accurate summary of DOD's use and field tests of remedial technologies; DOD also provided technical clarifications that we have incorporated, as appropriate.

We are sending copies of this report to appropriate congressional committees; the Secretary of Defense; the Administrator of EPA; and other interested parties. We will also make copies available to others upon request. In addition, the report will be available at no charge on GAO's Web site at http://www.gao.gov.

If you or your staff have any questions about this report, please contact me at (202) 512-3841 or mittala@gao.gov. Contact points for our Offices of Congressional Relations and Public Affairs may be found on the last page of this report. GAO staff who made major contributions to this report are listed in appendix V.

Anu K. Mittal
Director, Natural Resources and Environment

APPENDIX I
OBJECTIVES, SCOPE, AND METHODOLOGY

This report (1) describes the groundwater remediation technologies that the Department of Defense (DOD) is currently using or field-testing and (2) examines whether any new groundwater remediation technologies are being used outside the department or are being developed by commercial vendors that may have potential for DOD's use, and the extent to which DOD is researching and developing new approaches to groundwater remediation. In addition, this report provides limited information on the key characteristics, benefits, and limitations of selected groundwater remediation technologies.

To address the first objective, we developed a questionnaire that we sent to the DOD components responsible for DOD's groundwater cleanup efforts—the Air Force, Army, U.S. Army Corps of Engineers, Defense Logistics Agency, and Navy. In the questionnaire, we listed groundwater remediation technologies and asked these DOD components to indicate which technologies they have implemented and still currently use. We also asked the components to provide examples of specific groundwater remediation projects. We developed the list of technologies based on a review of reports and existing lists developed by the National Research Council, Environmental Protection Agency (EPA), Federal Remediation Technology Roundtable, and others, as well as through discussions with a groundwater remediation consulting firm and several nationally recognized groundwater remediation experts. To better understand DOD's processes for environmental cleanup and technology development, we met with officials from the offices of the Deputy Undersecretaries of Defense for Installations and Environment and for Science and Technology. We also reviewed documents, reports, and guidance on groundwater remediation from the Office of the Secretary of Defense and the various DOD components involved in groundwater remediation. To obtain information on how DOD uses groundwater remediation technologies to treat contaminants of concern, we toured several bioremediation projects at Dover Air Force Base and spoke with a groundwater remediation program manager for the Air Force.

To address our second objective, we contracted with consultants from the Washington, D.C., office of Malcolm Pirnie Inc. to gather information from commercial vendors on the range of currently available groundwater remediation technologies. We also attended a national groundwater remediation conference, where we spoke with a number of vendors of groundwater remediation technologies about their products, efforts to develop innovative approaches to groundwater remediation, and remediation work they may have performed for DOD. In addition, we collected and reviewed reports and studies from these vendors to better understand the range of technologies available to DOD. We also consulted with four nationally recognized groundwater remediation experts—two from academia and two from industry—to provide information on innovative remediation technologies currently available or under development by the commercial sector. We selected these experts on the basis of their independence, knowledge of and experience with groundwater remediation technologies, and recommendations from the National Academy of Sciences and others. In addition, we consulted with a senior groundwater remediation official from EPA's Groundwater and Ecosystem Restoration Division, who is an expert on technologies used for groundwater remediation.

Through these sources, we identified 15 technologies that are currently available commercially for the treatment of contaminated groundwater. For the purposes of this report, we defined a technology as a distinct technical method or approach for treating or removing contaminants found in groundwater. We did not consider any modifications or enhancements to a technology, such as variations in the material or equipment used during treatment, to be a separate technology. To determine whether there were any technologies currently being used outside of DOD, we compared the list of 15 currently available technologies with information provided to us by DOD officials on technologies currently used by DOD for groundwater remediation.

To identify the extent to which DOD supports the research and development of new approaches to groundwater remediation, we interviewed officials from the Strategic Environmental Research and Development Program and the Environmental Security Technology Certification Program. We reviewed reports, project portfolios, and other documents developed by these two programs. To gain a better understanding of DOD's efforts to field-test innovative approaches to groundwater remediation, we visited a DOD National Environmental Technology Test Site, located in Delaware, where private and public researchers can test innovative groundwater remediation technologies. We observed several ongoing research projects and interviewed an official responsible for managing the test facility. To gain a better understanding of DOD's relationship with the Interstate Technology and Regulatory Council, we reviewed a memorandum of understanding between the two organizations and interviewed an official that serves as DOD's liaison to the council.

Information presented in this report is based on publicly available documents and information provided by government officials, independent consultants, and experts. We did not review nonpublic research and development activities that may be under way in private laboratories. We reviewed data for accuracy and consistency, and corroborated DOD-provided data to the extent possible. We assessed the reliability of the DOD-provided data by reviewing related documentation, including DOD's annual reports to Congress on its Defense Environmental Restoration Program and information provided by consultants.

We performed our work from January 2005 through May 2005, in accordance with generally accepted government auditing standards.

APPENDIX II
TECHNOLOGIES FOR THE REMEDIATION OF CONTAMINATED GROUNDWATER

Ex-Situ Technologies

1. *Advanced oxidation processes* often use ultraviolet light irradiation with oxidizers such as ozone or hydrogen peroxide to produce free radicals, which break down and destroy chlorinated solvents, fuels, and explosive contaminants as water flows through a treatment reactor tank. Depending on the design of the system, the final products of this treatment can be carbon dioxide, water, and salts. An advantage of advanced oxidation processes is that it destroys the contaminant, unlike some other technologies, which only shift the phase of the contaminant into something more easily handled and removed. There are some limitations to these processes; for instance, maintenance of the treatment equipment can be a problem if certain substances—such as insoluble oil or grease—are allowed into the system. Also, the handling and storage of oxidizers can require special safety precautions. The cost of this type of remediation is largely dependent on the volume and flow rate of groundwater to be treated, energy requirements, and chemicals utilized. Operations and maintenance costs are also a factor in the overall cost of this approach. For the purposes of this report, advanced oxidation processes also include the related technologies of phyotolysis and photocatalysis.

2. *Air stripping* involves the mass transfer of volatile contaminants from water to air by exposing contaminated water to large volumes of air, so that the contaminants, such as chemical solvents, undergo a physical transformation from liquid to vapor. In a typical air stripper setup, called a packed tower, a spray nozzle at the top of a tower pours contaminated water over packing media or perforated trays within the tower. At the bottom of the tower, a fan forces air up through the tower countercurrent to the water flow, thus stripping the contaminants from the water. The contaminants in the air leaving the tower must then be removed and disposed of properly. Air strippers can be combined with other technologies for treatment of groundwater. Advantages of this technology include its potential to effectively remove the majority of the volatile organic contaminants of concern. Moreover, this mature technology is relatively simple and design practices are standardized and well-documented, and, in

comparison with other approaches, this technology is often less expensive. However, maintenance can be an issue with this technology if inorganic or biological material clogs or fouls the equipment, and process energy costs can be high.

3. *Bioreactors* are biochemical-processing systems designed to degrade contaminants in groundwater using microorganisms, through a process similar to that used at a conventional wastewater treatment facility. Contaminated groundwater flows into a tank or basin, where it interacts with microorganisms that grow and reproduce while degrading the contaminant. The excess biomass produced is then separated from the treated water and disposed of as a biosolids waste. This technology can be used to treat, among other things, chlorinated solvents, propellants, and fuels. Potential advantages of bioreactors include relatively low operations and maintenance costs and the destruction, rather than mass transfer of, the contaminants. Moreover, regulators and other stakeholders generally accept bioreactor technology as a proven approach for remediation. Nonetheless, there are some limitations to the use of bioreactors, including decreases in effectiveness if contaminant concentrations in the influent water are too high or too low to support microorganism growth and if nuisance microorganisms enter the system. Additionally, the sludge produced at the end of the process may need further treatment or specialized disposal. Bioreactor cost is influenced by the upfront capital needed for installation, setup, and start-up, as well as the operations and maintenance costs associated with longer-term treatment.

4. *Constructed wetlands* use artificial wetland ecosystems (organic materials, microbial fauna, and algae) to remove metals, explosives, and other contaminants from inflowing water. The contaminated water flows into the wetland and is processed by wetland plants and microorganisms to break down and remove the contaminants. Wetlands, intended to be a long-term remediation approach, can be created with readily available equipment and generally can operate with low maintenance costs. Furthermore, because this technology provides a new ecosystem for plant and animal life, it is generally popular with the public. However, this approach is often more suitable for groundwater that is ultimately discharged to the surface rather than reinjected into the ground. Also, the long-term effectiveness of this treatment is not well-known, as aging wetlands may lose their ability to process certain contaminants over time. Temperature, climate, and water flow rate may negatively impact the processes that break down the contaminants. Applicability and costs associated with constructed wetlands vary depending on site conditions, such as groundwater flow rate, contaminant properties, landscape, topography, soil permeability, and climate.

5. *Ion exchange* involves passing contaminated water through a bed of resin media or membrane (specific to the particular contaminant) that exchanges ions in the contaminants' molecular structure, thus neutralizing them. This approach can be useful for dissolved metals (e.g., hexavalent chromium) and can be used to treat propellants such as perchlorate. Once the ion exchange resin has been filled to capacity, it can be cleaned and reused (following a process called resin regeneration). Ion exchange is usually a short- to medium-term remediation technology. This technology allows contaminated water to be treated at a high flow rate and can completely remove the contaminants from the water. However, some substances—such as oxidants or suspended solids—in the incoming water may diminish the effectiveness of the ion exchange resins. Furthermore, different resin types can be

needed for different contaminants. Among the factors influencing costs are discharge requirements, the volume of water to be treated, contaminant concentration (as well as the presence of other contaminants), and resin regeneration. For the purposes of this report, ion exchange includes technologies that use ion exchange resins or reverse osmosis membranes to remove contaminants from groundwater, including dissolved metals and nitrates.

6. *Adsorption (mass transfer)* technologies involve passing contaminated water through a sorbent material—such as activated carbon—that will capture the contaminants (through either adsorption or absorption), thus removing or lessening the level of contaminants in the water. The contaminated water is pumped from the aquifer and passed through the treatment vessel containing the sorbent material. As the contaminated water comes into contact with the sorbent surface, it attaches itself to that surface and is removed from the water. Benefits of this technology include its ability to treat contaminated water to nondetectable levels and its potential for treating low to high groundwater flow rates as well as multiple contaminants simultaneously. However, some contaminants may not be sorbed well or the sorbent unit may require disposal as hazardous waste. Furthermore, this approach is impractical if the contaminant levels are high due to higher costs resulting from frequent changing of the sorbent unit. If the concentrations of contaminants are low or flow rates for treatment can be kept low, then adsorption technology may be a cost-effective approach.

In-Situ Technologies

1. *Air sparging* introduces air or other gases into a contaminated aquifer to reduce concentrations of contaminants such as fuel or chlorinated solvents. The injected air creates an underground air stripper that removes contaminants by volatilization (a process similar to evaporation that converts a liquid or solid into a gas or vapor). This injected air helps to transport the contaminants up into the unsaturated zone (the soil above the water table, where pores are partially filled with air), where a soil vapor extraction system is usually implemented to collect the vapors produced through this process. This technology has the added benefit of often stimulating aerobic biodegradation (bioremediation) of certain contaminants because of the increased amount of oxygen introduced into the subsurface. Typically, air sparging equipment is readily available and easily installed with minimal disturbance to site operations. However, this technology cannot be used if the contaminated site contains contaminants that don't vaporize or are not biodegradable. In some cases, this technology may not be suitable for sites with free product (e.g., a pool of fuel floating on the water table) because air sparging may cause the free product to migrate and spread contamination. Also, this technology is less effective in highly stratified or heterogeneous soils since injected air tends to travel along paths of least resistance in the subsurface, potentially bypassing areas of contamination. This technology can be less costly than ex-situ technologies because it does not require the removal, treatment, storage, or discharge of groundwater. For the purposes of this

report, air sparging includes the related remedial approaches of co-metabolic sparging, sparging using other gases, and in-well air stripping.

2. *Bioremediation* relies on microorganisms to biologically degrade groundwater contaminants through a process called biodegradation. It may be engineered and accomplished in two general ways: (1) stimulating native microorganisms by adding nutrients, oxygen, or other electron acceptors (a process called biostimulation); or (2) providing supplementary pregrown microorganisms to the contaminated site to augment naturally occurring microorganisms (a process called bioaugmentation). This technology mainly focuses on remediating organic chemicals such as fuels and chlorinated solvents. One approach, aerobic bioremediation, involves the delivery of oxygen (and potentially other nutrients) to the aquifer to help native microorganisms reproduce and degrade the contaminant. Another approach, anaerobic bioremediation, circulates electron donor materials—for example, food-grade carbohydrates such as edible oils, molasses, lactic acid, and cheese whey—in the absence of oxygen throughout the contaminated zone to stimulate microorganisms to consume the contaminant. In some cases, pregrown microbes may be injected into the contaminated area to help supplement existing microorganisms and enhance the degradation of the contaminant, a process known as bioaugmentation. A potential advantage of bioremediation is its ability to treat the contaminated groundwater in place with naturally occurring microorganisms, rather than bringing contaminants to the surface. By using native microorganisms, rather than injecting additional ones, cleanup can be more cost-effective at some sites. However, heterogeneous subsurfaces can make delivering nutrient/oxygen solutions to the contaminated zone difficult by trapping or affecting movement of both contaminants and groundwater.[1lso, nutrients to stimulate the microorganisms can be consumed rapidly near the injection well, thereby limiting the microorganisms' contact with the contaminants, or stimulating biological growth at the injection site. In summary, this technology avoids the costs associated with bringing water to the surface for treatment; instead, the main costs associated with bioremediation include: delivery of the amendments to the subsurface (which varies depending on the depth of contamination), the cost of the amendments themselves, and monitoring of the treatment. For the purposes of this report, bioremediation includes the related bioremedial approaches of bioaugmentation, biostimulation, co-metabolic treatment, enhanced aerobic biodegradation, enhanced anaerobic biodegradation, and biobarriers.

3. *Enhanced recovery using surfactant flushing* speeds contaminant removal in conventional pump-and-treat systems by injecting surfactants[2nto contaminated aquifers or soil to flush the contaminant toward a pump in the subsurface (some distance away from the injection point); this pump removes the contaminated water and surfactant solution to the surface for treatment and disposal of contaminants. Surfactants are substances that associate with organic compounds such as fuels and chlorinated solvents and significantly increase their solubility, which aids cleanup of contaminated aquifers with less flushing water and pumping time. This technology is applicable to both dense and light nonaqueous phase liquids (DNAPL and LNAPL).[3enefits of enhanced recovery approaches include the rapid removal of contaminants, which may significantly reduce cleanup times. However, regulatory

issues may require special attention due to extra scrutiny for obtaining approvals to inject surfactant solutions; a greater degree of site characterization is often required to satisfy both technical and regulatory requirements. In addition, subsurface heterogeneities and low permeability can interfere with the effective delivery and recovery of the surfactant solution. Furthermore, to the extent that mobilization of organic liquid contaminants is achieved, this approach may be better for LNAPLs than DNAPLs, as LNAPLs tend to migrate upward and DNAPLs downward, possibly trapping them in previously uncontaminated subsurface areas. In addition to the high cost of surfactant solutions, another factor influencing the overall cost of this approach may be the treatment of the surfactant solution that is pumped out of the aquifer. For the purposes of this report, this technology includes related remedial approaches that use co-solvents such as ethanol to improve the solubility of surfactants in the subsurface.

4. *Chemical treatments* include remediation technologies that chemically oxidize or reduce contaminants when reactive chemicals are injected into the groundwater. This approach converts contaminants such as fuels and explosives into nonhazardous or less-toxic compounds. Depending on the extent of contamination, this process involves injecting chemicals into the groundwater and generally takes a few days to a few months to observe results in rapid and extensive reactions with various contaminants of concern. Additionally, this technology can be tailored to the site and does not require rare or complex equipment, which may help reduce costs. Generally, there are no unusual operations and maintenance costs; however, in-situ chemical treatment may require intensive capital investment for large contaminant plumes or zones where repeated applications or large volumes of reactive chemicals may be required; major costs are associated with injection-well installation (cost influenced by well depth), procurement of the reactive chemicals, and monitoring. Additionally, site characterization is important for the effective delivery of reactive chemicals, as subsurface heterogeneities may result in uneven distribution of the reactive chemicals. For the purposes of this report, chemical treatment also includes various remedial approaches and technologies that chemically oxidize or reduce contaminants in-situ, as well as those that result in the in-situ immobilization and stabilization of soluble metals.

5. *Monitored natural attenuation* is a relatively passive strategy for in-situ remediation that relies on the naturally occurring physical, chemical, and biological processes that can lessen concentrations of certain contaminants in groundwater sufficiently to protect human health and the environment. The changes in contaminant concentrations are observed through various wells that are placed throughout the contaminated groundwater zone to monitor the level of contamination over time and its migration from its initial location in the subsurface. Some chlorinated solvents and explosives may be resistant to natural attenuation; however, it can still be used in cases of nonhalogenated chlorinated solvents and some inorganic compounds. If appropriate for a given site, natural attenuation can often be less costly than other forms of remediation because it requires less infrastructure, construction, and maintenance. Furthermore, it is less intrusive because fewer surface structures are necessary and it may be used in all or selected parts of a contaminated site, alone or in conjunction with other types of remediation. However, compared with active

techniques, natural attenuation often requires longer time frames to achieve remediation objectives.

6. *Multiphase extraction* uses a series of pumps and vacuums to remove free product,[4ontaminated groundwater, and vapors from the subsurface, treat them, and then either dispose or reinject the treated groundwater. Specifically, one or more vacuum extraction wells are installed at the contaminated site to simultaneously pull liquid and gas from the groundwater and unsaturated soil directly above it. This type of vacuum extraction well removes contaminants from above and below the groundwater table, and can expose more of the subsurface for treatment, notably in low permeability or heterogeneous formations. The contaminant vapors are collected in the extraction wells and taken above ground for treatment. This approach can be used to treat organic contaminants—such as chlorinated solvents and fuels—and can be combined with other technologies, particularly above-ground liquid/vapor treatment, as well as other methods of in-situ remediation such as bioremediation, air sparging, or bioventing. Potential advantages of this technology include its applicability to groundwater cleanup in low permeability and heterogeneous formations and its minimal disturbance to site-specific conditions. However, the system requires complex monitoring and specialized equipment, and it may be difficult or problematic to implement the most effective number of pumps. A major contributor to this technology's cost is operations and maintenance, which may run from 6 months to 5 years, depending on site-specific factors. For the purposes of this report, multiphase extraction includes the related technologies of bioslurping and dual-phase extraction.

7. *Permeable reactive barriers* are vertical walls or trenches built into the subsurface that contain a reactive material to intercept and remediate a contaminant plume as the groundwater passes through the barrier. This technology can be used to treat a wide range of contaminants and is commonly used to treat chlorinated solvents and heavy metals. Reactive barriers usually do not require above-ground structures or treatment, allowing the site to be used while it is being treated. However, its use is limited by the size of the plume since larger contaminant plumes are often more difficult to intercept for treatment. Moreover, the barrier may lose effectiveness over time as microorganisms or chemicals build up on the barrier, making rehabilitation or media replacement necessary. The depth of the contaminated groundwater zone and the required barrier may also present some technical challenges. Underground utility lines, rocks, or other obstacles can increase the difficulty of installing a barrier and drive up capital costs. Additionally, because permeable reactive barriers do not treat the contaminant source, but simply the plume, treatment may be required for extended time periods, thus increasing overall cleanup costs. For the purposes of this report, permeable reactive barriers include biotic and abiotic, as well as passive and active treatment barriers.

8. *Phytoremediation* is the use of selected vegetation to reduce, remove, and contain the toxicity of environmental contaminants, such as metals and chlorinated solvents. There are several approaches to phytoremediation that rely on different plant system processes and interactions with groundwater and contaminants. One approach to phytoremediation is phytostabilization, which uses plants to reduce contaminant mobility by binding contaminants into the soil or incorporating contaminants into

plant roots. Another approach is phytoaccumulation, where specific species of plants are used to absorb unusually large amounts of metals from the soil; the plants are later harvested from the growing area and disposed of in an approved manner. A similar process is called rhizofiltration, where contaminated water moves into mature root systems and is circulated through their water supply. Another process can remove contaminants by evaporating or volatilizing the contaminants from the leaf surface once it has traveled through the plant's system. Phytoremediation offers the benefit of only minimally disturbing the environment and can be used for the treatment of a wide range of contaminants. However, specific plant species required for particular contaminants may be unable to adapt to site conditions due to weather and climate, and phytoremediation may not be an effective approach for deep contamination. While maintenance costs, including cultivation, harvesting, and disposal of the plants, are substantial for this technology, phytoremediation typically has lower costs than alternative approaches. For the purposes of this report, phytoremediation includes phytostabilization, phytoaccumulation, phytoextraction, rhizofiltration, phytodegredation, rhizosphere degredation, organic pumps, and phytovolitilization.

9. *Thermal treatments* involves either pumping steam into the aquifer or heating groundwater in order to vaporize chlorinated solvents or fuels from the groundwater. The vaporized contaminant then rises into the unsaturated zone and can be removed via vacuum extraction for treatment. There are three main approaches for heating the groundwater in-situ. The first, radio frequency heating, uses the electromagnetic energy found in radio frequencies to rapidly heat the soil in a process analogous to microwave cooking. The second, electromagnetic heating, uses an alternating current to heat the soil and may include hot water or steam flushing to mobilize contaminants. The third uses heating elements in wells to heat the soil. Thermal treatments may be applied to a wide range of organic contaminants and sites with larger volumes of LNAPLs or DNAPLs as well as sites with low permeability and heterogeneous formations. However, the presence of metal and subsurface heterogeneities in the contaminated site may interfere with this process. The heating and vapor collection systems must be designed and operated to contain mobilized contaminants, to avoid their spread to clean areas. The major costs incurred for thermal treatments are for moving specialized equipment to the site, developing infrastructure to provide power, and providing energy to run the system. For the purposes of this report, thermal treatments include related soil-heating technologies, such as steam flushing, conductive heating, and electrical resistance heating.

APPENDIX III
GROUNDWATER REMEDIATION
EXPERTS CONSULTED

Dr. John Fountain
Professor and Head, Department of Marine, Earth and
Atmospheric Sciences

North Carolina State University
Raleigh, North Carolina

Dr. Robert E. Hinchee
Principal Civil and Environmental Engineer
Integrated Science and Technology Inc.
Panacea, Florida

Dr. Michael C. Kavanaugh
Vice President
National Science and Technology Leader
Malcolm Pirnie Inc.
Emeryville, California

Dr. Robert L. Siegrist
Professor and Division Director
Environmental Science and Engineering Division
Colorado School of Mines
Golden, Colorado

Dr. John T. Wilson
Senior Research Microbiologist
Ground Water and Ecosystems Restoration Division, National Risk
Management Research Laboratory
Office of Research and Development
U.S. Environmental Protection Agency
Ada, Oklahoma

REFERENCES

[1]   Remediation of a contaminated site involves efforts to remove, destroy, or isolate contaminants found in the groundwater. In some cases, disposal practices at these sites predate the enactment of relevant environmental cleanup statutes.

[2]   The Navy oversees environmental restoration on Marine Corps facilities.

[3]   The Corps may also participate in groundwater remediation activities on active Army installations, some Air Force installations, and properties that are scheduled for closure as part of the Base Realignment and Closure Act process.

[4]   For the purposes of this report, we have defined a "site" as a specific area of contamination and a "facility" as a geographically contiguous area under DOD's ownership or control within which a contaminated site or sites are located. A single DOD facility may contain multiple sites requiring cleanup.

[5]   Ronald W. Reagan National Defense Authorization Act for Fiscal Year 2000, Pub. L. No. 108-375, § 316, 118 Stat. 1811, 1843 (Oct. 28, 2004).

[6]   See appendix II for more information on each of the 15 technologies.

[7] DOD carries out some groundwater remediation as corrective action under the Resource Conservation and Recovery Act of 1976 (RCRA). According to DOD, while RCRA and CERCLA contain somewhat different procedural requirements, these differences do not substantively affect the outcome of remedial activities.

[8] This list represents EPA's highest priorities for cleanup nationwide, including public and private sites considered by EPA to present the most serious threats to human health and the environment. To make its determination, EPA uses a hazard-ranking system to evaluate the severity of the contamination by examining the nature of the contaminants, the pathways through which they can move (such as soil, water, or air), and the likelihood that they may come into contact with a receptor—for example, a person living nearby. According to DOD's

[9] Defense Environmental Programs, Annual Report to Congress, Fiscal Year 2004, DOD has 152 facilities that are listed or proposed for listing on the National Priorities List.

[10] 42 U.S.C. § 9621(c). The applicable EPA regulation differs from the statute: It requires the five-year reports only if contaminants will remain at the site "above levels that allow for unlimited use and unrestricted exposure." 40 C.F.R. § 300.430(f)(4)(ii).

[11] Surfactants, or surface active agents, are molecules with two structural units: one with an affinity for water and one with an aversion to water. This molecular combination is useful for dissolving some contaminants and enhancing their mobility by lowering the interfacial tension between the contaminant and the water.

[12] For more information, see National Research Council, Water Science and Technology Board, Contaminants in the Subsurface: Source Zone Assessment and Remediation (Washington, D.C., 2004).

[13] A contaminant may exist in aqueous (dissolved in water), nonaqueous, solid (sorbed), or gaseous form.

[14] For more information, see National Research Council, Contaminants in the Subsurface: Source Zone Assessment and Remediation (Washington, D.C., 2005).

[15] For more information, see EPA, Office of Solid Waste and Emergency Response, Remediation Technology Cost Compendium—Year 2000 (Washington, D.C., 2001).

[16] For additional information, see the online version of the Federal Remediation Technologies Roundtable Treatment Technologies Screening Matrix at http://www.frtr.gov/scrntools.htm.

[17] Nanoscale refers to miniscule particles that measure less than 100 nanometers in diameter. In comparison, an average human hair typically measures 10,000 nanometers in diameter.

[18] See National Research Council, Water Science and Technology Board, Environmental Cleanup at Naval Facilities: Adaptive Site Management (Washington, D.C., 2003).

[19] SERDP's goals include supporting basic and applied research and development of environmental technologies; providing information and data on environmental research and development activities to other governmental and private organizations in an effort to promote the transfer of innovative technologies; and identifying technologies developed by the private sector that are useful for DOD's and DOE's environmental restoration activities.

[20] According to ESTCP, the program "provides an independent, unbiased evaluation of the cost, performance, and market potential of state-of-the-art environmental

technologies based on field demonstrations conducted under DOD operational conditions."

APPENDIX II

[1] Heterogeneities can cause wide variability in hydraulic properties such as hydraulic conductivity—a measure of the volume of water that will pass through an area at a given time. These changes in hydraulic properties enhance the dispersion of a dissolved contaminant spread. Heterogeneities can also create preferential pathways for contaminant migration.

[2] Surfactants are molecules with two structural units: one with an affinity for water and one with an aversion to water. Surfactants are especially useful for dissolving some contaminants and enhancing their mobility by lowering the interfacial tension between the contaminant and water.

[3] Nonaqueous-phase liquids are liquids that do not mix with, or dissolve in, water. Dense nonaqueous-phase liquids (DNAPL) fall to the bottom of a body of water; chlorinated solvents are typical examples. Conversely, light nonaqueous-phase liquids (LNAPL) gather on top of the water. Gasoline and fuel oil are examples of LNAPLs.

[4] Free products are liquid contaminants floating on top of groundwater.

In: Groundwater Research and Issues   ISBN: 978-1-60456-230-9
Editors: W. B. Porter, C. E. Bennington, pp. 179-184   © 2008 Nova Science Publishers, Inc.

Chapter 7

THE IMPACTS ON WATER QUALITY FROM PLACEMENT OF COAL COMBUSTION WASTE IN PENNSYLVANIA COAL MINES

Clean Air Task Force

COAL COMBUSTION WASTE DISPOSAL IN PENNSYLVANIA

In a multi-year study, the Clean Air Task Force (CATF) examined 15 coal mines where coal ash was placed under the Pennsylvania Department of Environmental Protection (PADEP) Coal Ash Beneficial Use Program, which encourages the placement of coal combustion waste (CCW) in active and abandoned mines. The study concludes that the state's beneficial use program, whose primary goal is to improve the environmental condition of mines by adding of massive quantities of CCW, is failing:

- At 10 of the 15 minefills examined in the study, monitoring data indicate the coal ash contaminated groundwater or streams.
- At three minefills contamination of streams and/or groundwater was occurring, but the cause of the pollution could not definitively be traced to the ash because of the lack of monitoring data.
- At one of the minefills, water quality improved for acid mine parameters, but water quality decreased for contaminants found in ash.
- At only 1 of the 15 minefills was water quality improvement found. However, since water monitoring was terminated shortly after placement, it is unknown whether the improvement was temporary.

Pennsylvania generates over 9 million tons of coal combustion waste a year, the third largest producer in the country.

Consequently, CATF found that placing large amounts of CCW in mines is a dangerous practice that appears to be causing toxic levels of contamination. The report recommends that permits allowing this industrial waste to be placed in mines require safeguards to minimize adverse Environmental impacts and threats to human health.

CCW is a toxic industrial waste produced by coal-burning power plants. Pennsylvania is the third largest US producer of this waste, generating over 9 million tons per year. CCW contains hazardous chemicals including aluminum, chloride, iron, manganese, sulfate and toxic trace elements such as arsenic, selenium, lead, mercury, cadmium, nickel, copper, chromium, boron, molybdenum and zinc.

For over 20 years, PADEP has been promoting placement of large volumes of CCW in active and abandoned coal mines to address acid mine drainage, increase soil fertility, and fill mine pits and voids. PADEP has permitted approximately 120 CCW minefills throughout the state.

THE CLEAN AIR TASK FORCE REPORT

The CATF report shows that placement of CCW in mines can produce significant pollution. For example:

- At the Ernest Mine where more than 1.5 million tons of CCW have been placed, water contaminated with lead (9.7 times the federal drinking water standard), cadmium (almost 15 times the standard) and chromium (2.4 times the standard) has been discharging into surface water.

- At the McDermott Mine where 316,000 tons of CCW were placed, cadmium and selenium have been measured 76 times and 36 times higher than water quality standards, respectively, at mine discharges. An offsite spring used as a domestic water supply had to be abandoned due to pollution from the site.
- At the Swamp Poodle Mine where 214,000 tons of CCW were placed, arsenic rose to 389 times the drinking water standard, cadmium to 46 times the standard, and lead and selenium to four to nearly seven times the standard in groundwater.
- At the Ellengowan and BD Mines, lead has been measured up to 39 times and cadmium up to 32 times federal drinking water standards in mine pools downgradient of four pits containing more than 16 million tons of CCW.

Recent congressional concern about CCW disposal in mines lead to a 2006 National Academies of Science (NAS) report entitled *Managing Coal Ash Residue in Mines*. The NAS concluded that certain safeguards, not required by the PA program, are essential to protect health and the environment, including adequate characterization of both the CCW and the mine site, adequate monitoring of the ash after placement, isolation from water, cleanup standards, and meaningful public participation in the permitting process. Furthermore, the NAS recommended that no placement of CCW in mines should occur if other safer reuse alternatives exist, such as incorporation of ash into concrete and other products.

US EPA has also expressed serious concern over CCW minefilling. In its 2000 Regulatory Determination on Wastes from the Combustion of Fossil Fuels, the Agency stated:

> "We are aware of situations where coal combustion wastes are being placed in direct contact with ground water in both underground and surface mines. This could lead to increased releases of hazardous metal constituents as a result of minefilling. Thus if the complexities related to site-specific geology, hydrology, and waste chemistry are not taken into account when minefilling coal combustion wastes, we believe that certain minefilling practices have the potential to degrade, rather than improve, existing groundwater quality and can pose a threat to human health and the environment."

US EPA concluded, and the NAS concurred, that enforceable federal regulations are necessary to guarantee that state programs minimize the threats from minefilling, but the Agency has not yet taken action to promulgate federal standards.

Coal combustion waste disposal sites in PA are contaminated with toxic levels of arsenic, cadmium, selenium and other pollutants.

RECOMMENDATIONS TO PADEP

CATF's report carefully examined the PA CCW minefilling program and found it lacking in critical respects, including the failure to recognize degradation from the use of CCW. The report recommends PADEP:

- Require accurate and thorough waste and site characterization prior to permitting the use of coal ash in mines.
- Integrate waste and site characterizations and update them as new information becomes known to ensure effective monitoring.
- Require comprehensive, long-term water quality monitoring at all coal ash mine placement sites.
- Include enforceable corrective action standards for coal ash parameters in all coal ash mine placement permits and address degradation that occurs.

- Issue NPDES permits that monitor and control ash contaminants in surface discharges from sites.
- Require financial assurance that addresses potential long-term water quality problems at coal ash mine placement sites.
- Require isolation of coal ash from groundwater at all coal ash placement sites.
- Update its permit system with a better organized more publicly accessible modern database.
- Require that all coal ash placement permits in mines actually achieve a measurable beneficial result.
- Establish enforceable requirements for coal ash placement permits in state regulations to replace the current system of unenforceable guidance documents.
- Establish a program to promote the safe reuse of coal ash, prior to issuing or renewing permits for coal ash minefills, and only if such safe and Beneficial recycling is unavailable, permit Placement of CCW in mines.

Coal combustion waste disposal in Pennsylvania has caused tremendous problems across the state. Currently there are approximately 120 CCW minefills in Pennsylvania.

WHERE IS CONTAMINATION OCCURING FROM THE PLACEMENT OF COAL COMBUSTION WASTE?

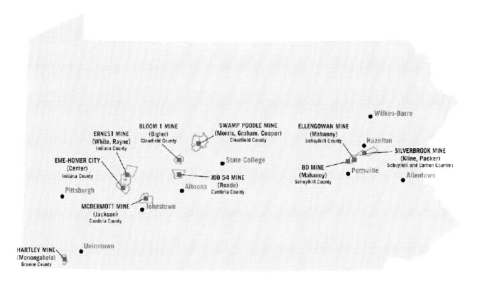

10 COAL ASH PLACEMENT SITES WHERE CONTAMINATION WAS FOUND

"... We believe that certain minefilling practices have the potential to degrade, rather than improve, existing groundwater quality and can pose a threat to human health and the environment."

-US EPA

"Thus the committee concludes that the presence of high contaminant levels in many CCR [coal combustion residue] leachates may create human health and ecological concerns at or near some mine sites over the long term."

-National Academies of Science

In: Groundwater Research and Issues   ISBN: 978-1-60456-230-9
Editors: W. B. Porter, C. E. Bennington, pp. 185-198   © 2008 Nova Science Publishers, Inc.

Chapter 8

MINERALOGICAL PRESERVATION OF SOLID SAMPLES COLLECTED FROM ANOXIC SUBSURFACE ENVIRONMENTS

*Richard T. Wilkin**

BACKGROUND

Remedial technologies utilized at hazardous waste sites for the treatment of metal and metalloid contaminants often take advantage of reduction-oxidation (redox) processes to reach ground water clean up goals (Barcelona and Holm, 1991; U.S. Environmental Protection Agency, 2002). This is because redox reactions, in many cases, govern the biogeochemical behavior of inorganic contaminants by affecting their solubility, reactivity, and bioavailability. Site characterization efforts, remedial investigations, and long-term post-remedial monitoring often involve sampling and analysis of solids. Solid-phase studies are needed to evaluate contaminant partitioning to various mineral fractions, to develop site conceptual models of contaminant transport and fate, and to assess whether or not remedial mechanisms are occurring as expected. Measurements to determine mineralogical compositions, contaminant-mineral associations, and metal/metalloid uptake capacities of subsurface solids or reactive media used for *in situ* treatment of the subsurface all depend upon proper sample collection and preservation practices. This Issue Paper discusses mineralogical preservation methods for solid samples that can be applied during site characterization studies and assessments of remedial performance. A preservation protocol is presented that is applicable to solids collected from anoxic subsurface environments, such as soils, aquifers, and sediments.

The preservation method evaluated and recommended here for solids collected from anoxic environments involves sample freezing (-18 °C), transportation of frozen samples on dry ice, and laboratory processing of solids in an anaerobic glove box. This method was

---

* U.S. Environmental Protection Agency, National Risk Management Research Laboratory, Ground Water and Ecosystems Restoration Division, 919 Kerr Research Drive, Ada, OK 74820 (wilkin.rick@epa.gov)

found to preserve the redox integrity of reduced iron- and sulfur-bearing compounds, which are typically predominant redox-sensitive inorganic constituents in environmental materials andareimportantincontrollingcontaminantbehaviorathazardous waste sites. A selection of solid-phase measurements was carried out on preserved anoxic sediments collected from a contaminated lake and compared to identical measurements on sample splits in which no preservation protocol was adopted, i.e., the unpreserved samples were allowed to oxidize in ambient air. An analysis of results illustrates the importance of proper sample preservation for obtaining meaningful solid-phase characterization. This paper provides remedial project managers and other state or private remediation managers and their technical support personnel with information necessary for preparing sampling plans to support site characterization, remedy selection, and post-remedial monitoring efforts.

U.S. Environmental Protection Agency, National Risk Management Research Laboratory, Ground Water and Ecosystems Restoration Division, 919 Kerr Research Drive, Ada, OK 74820 (wilkin.rick@epa.gov). For further information contact Richard T. Wilkin at (580) 436-8874.

INTRODUCTION

Solid phase samples may be collected for physical, chemical, or biological tests during site characterization and remedial performance monitoring studies. The principal objective of any sampling program is to collect and deliver materials to the laboratory that are representative of the original material present in the environment. If samples are collected for the purpose of determining total element concentrations, then the mode of preservation may not be important unless the contaminant is a volatile or semi-volatile component. However, when solid samples are collected for more sensitive or detailed analyses, such as sequential extraction tests, solid-phasespeciationtests,orbatchadsorptiontests,preservation methods become critical and may direct the outcome of all subsequent analyses and interpretations. For samples collected from anoxic subsurface environments, oxidation is the primary reaction process that leads to unrepresentative samples. Therefore, proper sample preservation will ideally minimize the undesirable effects of oxidation. Unfortunately, the literature is not extensive on the assessment of procedures for handling anoxic materials. Lacking generalguidance,sampling andpreservationprotocolsare usually developed to best suit needs on a project-by-project basis.

Redox-sensitive elements commonly important in environmental studies include iron, manganese, sulfur, chromium, copper, uranium, and arsenic (U.S. Environmental Protection Agency, 2002). Reduction-oxidation processes involving iron and sulfur compounds, in particular, have significant impacts on the partitioning of metals to solids and these impacts must be considered when collecting and preserving field samples. For example, minerals containing ferrous iron (e.g., siderite, $FeCO_3$; mackinawite, FeS; pyrite, FeS_2) may undergo rapid oxidation reactions during air exposure and transform to ferric-iron phases (e.g., ferrihydrite, $Fe(OH)_3 \cdot nH_2O$; lepidocrocite, γ-FeOOH; goethite, α-FeOOH). Subsequently during batch adsorption tests or sequential extraction tests, ferric-bearing phases should behave differently than the original, unoxidized material representative of the natural environment. Oxidative mineral transformations may result in changes in reactive surface area, influence precipitation and co-precipitation reactions, and/or trigger different surface adsorption reactions. Similarly, sulfide minerals are in general highly susceptible to oxidation. Unpreserved samples containing sulfides can undergo oxidativetransformations, changingsamplebehaviorandoutcomes of mineralogical, sequential extraction, and batch adsorption tests (Bush and Sullivan, 1997;Carbonaro et al., 2005). A good example of how sample preservation practices can affect the outcome of sediment analyses is discussed in Harrington et al. (1999).

Adequate preservation of samples must encompass the chain of events from sample collection, to sample transport back to the laboratory, to sample storage, and finally to sample preparation and analysis in the laboratory. The strategies most often adopted for preserving the redox status of freshly collected solid materials include sample freezing and/or sample storage in an inert atmosphere. Sample freezing typically involves collection of core or grab samples, placement of samples in containers, followed by freezinginafreezerorflash-freezingusingliquidnitrogen. Freezing preserves the redox integrity of samples by decreasing the rate of reaction between reduced solids and atmospheric oxygen or other oxidants. The other principal approach for preserving redox status is to eliminate or minimize sample interactions with oxygen by transferring samples after their collection into an evacuated container or a container purged with an inert gas, such as nitrogen, helium, or argon. Transportation of samples back to the laboratory is an especially vulnerable process for maintaining sample integrity. Once samples are frozen, they must be kept in a frozen state, or samples stored in gas-purged containers must be transported in secondary air-tight containers. After samples have arrived at the laboratory, they may be again transferred to a laboratory freezer or to an anaerobic chamber for drying and homogenization.

Givelet et al. (2004) recently developed a protocol for the collection and handling of peat samples for chemical and mineralogical analyses. Their protocol adopts sample freezing at -18 °C to preserve samples for subsequent mineralogical analyses. Rapinetal.(1986) examined the impact of freezing and other preparation methods on the results of sequential extraction analyses for determining solid-phase partitioning of metals in sediment. Their conclusions were that freeze-drying and oven-drying should be avoided, but that freezing was acceptable for sequential extraction tests. They also noted that partial extraction tests for copper, iron, and zinc were especially sensitive to sample handling protocols. Mudroch and Bourbonniere (1994) proposed that when applying sequential extraction procedures to anoxic sediments all manipulations and extraction steps should be carried out in an anaerobic glove box. Mineralogical studies on the corrosion products in zero-valent iron permeable reactive barriers (PRBs) were conducted by Phillips et al. (2003). After collecting cores from subsurface PRBs, these investigators placed the core materials in PVC chambers purged with argon gas. The cores were stored for up to 2 weeks and argon was recharged into the PVC tubes every 2 to 3 days. Other long-term performance studies of zero-valent iron PRBs have successfully utilized sample freezing to preserve core materials for mineralogical and chemical analyses (U.S. Environmental Protection Agency, 2003).

Several studies have examined the effects of air-drying versus oven-drying on the behavior of soils and sediments in batch adsorption tests. Physical and chemical properties of materials are altered depending on the mode of drying. For example, differences have been observed in sample pH, partition coefficients, and the exchangeable metals fraction depending on whether samples are field-moist, frozen, air-dried, freeze-dried, or oven-dried. Based upon a review of previous work, U.S. Environmental Protection Agency(1992

)recommendsair-dryingofsamplesoveroven-drying in order to minimize changes to the physico-chemical properties of solids used in batch tests for estimating adsorption parameters. The endpoint of air-drying is achieved when the sample moisture content reaches equilibrium with room atmosphere conditions and in practice can be assessed by tracking sample mass to a steady state. Reduced solids collected from anoxic environments should be dried in an anaerobic glove box or glove bag to prevent oxidation (U.S. Environmental Protection Agency, 1992). The analysis of solid materials for inorganic species can be performed on wet, freeze-dried, or air-dried samples. In general, however, sample drying is preferred to eliminate sample homogeneity issues in relating element concentrations from a wet-weight to a dry-weight basis (Muhaya et al., 1998). Water removal may be achieved through various means, including decanting, gravity filtration, vacuum filtration, pressure filtration (e.g., Bottcher et al., 1997), and centrifugation. It is important to note that pore-water solutes can significantly contribute to total element concentrations in dried solids,especiallyin situationswherethe solid-phase concentration of the element is low (<10 mg/kg) and the pore-water concentration is high (>1 mg/L).

METHODS, RESULTS AND DISCUSSION OF A PRESERVATION STUDY

In order to evaluate the effects of sample preservation on the results of selected solid-phase characterization tests, contaminated sediments were collected from a small lake situated adjacent to a Superfund Site located approximately 16km northwest of downtown Boston, Massachusetts (Industri-Plex Superfund Site). The lake receives discharge of ground water with elevated concentrations of arsenic, ferrous iron, sulfate, and petroleum hydrocarbons. The site has been used to develop an improved understanding of arsenic geochemical cycling at the ground water-surface water interface (U.S. Environmental Protection Agency, 2005). Sediments were retrieved from depths ranging from 0.5 to 4.5 meters using an Eckman dredge. One half of each sample retrieved from the lake bottom was immediately bagged and frozen; the other half was bagged and left unfrozen. During each sampling event approximately 1 L of sediment plus water was collected. The mixture was transferred from the dredge to polyethylene bags and excess air was displaced. Frozen sediment samples were transported back to the laboratory on dry ice. Frozen samples were subsequently thawed and dried at room temperature in an anaerobic glove box (96:4 v/v N_2-H_2 gas mixture). The dried sediments were homogenized with an agate mortar and pestle and kept in the glove box. Unpreserved samples were dried in air and homogenized using an agate mortar and pestle. The color of the unpreserved samples was red, presumably due to the oxidation of ferrous iron and production of ferric oxyhydroxides. Color changes in the unpreserved samples were noted within the first several hours after sample collection. The preserved samples kept in the glove box remained black in color. Solid-phase tests carried out on the preserved and unpreserved sediments included total metals concentrations, metal extractability with 1 M HCl, total sulfur, acid-volatile sulfide, chromium-reducible sulfur, and batch adsorption tests with arsenic and zinc. In addition, X-ray absorption near-edge structure (XANES) spectroscopy was carried out to determine the oxidation state of arsenic in the preserved and unpreserved samples.

IRON AND SULFUR PARTITIONING

Total element concentrations were determined by microwave assisted digestion in nitric acid followed by inductively coupled plasma-optical emission spectroscopy (ICP-OES; modified EPA Method 3051). Figure 1 is a bar graph that shows a comparison of totalironconcentrationsinthepreservedandunpreservedsediment samples. Concentrations of total iron in the sediments range from
1.0 to 11.5 wt%. Total iron concentrations are independent of the mode of preservation; values in the preserved and unpreserved samples deviate within ±10%. Similar correlations are observed for other major and trace elements. As a general rule, therefore, the total concentration of inorganic components is conservative and independent of the mode of sample preservation. If total concentrations in solid samples are the data objective of a specific site investigation, then it may not be necessary to expend the extra effort and cost to ensure preservation of the sample redox state. Sample preservation may be necessary, however, to maintain solid-phase concentrations of volatile or semi-volatile inorganic components, such as mercury (Muhaya et al., 1998).

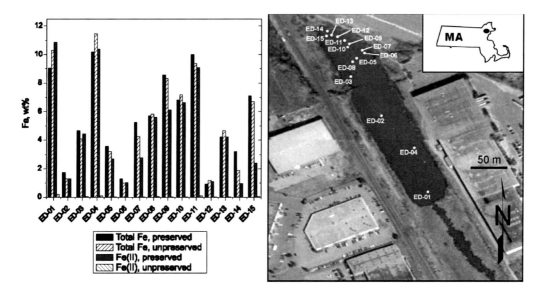

Figure 1. Comparison of solid-phase concentrations of total iron and ferrous iron (wt%) in a series of sediments with and without preservation. Map showing the distribution of sampling points within the Hall's Brook Holding Area pond, located adjacent to the Industri-Plex Superfund Site (for site background see U.S. Environmental Protection Agency, 2005 and references therein).

Although total concentrations of iron are independent of the mode of sample preservation, the oxidation state of iron in the preserved and unpreserved samples is completely different. Figure 1 shows the amount of ferrous iron in the solid phase compared to total iron concentrations in the preserved versus unpreserved sediments. Ferrous iron content was determined by extracting the sediments in 1 M HCl and measuring the ferrous iron concentration using the 1,10-phenanthroline colorimetric method. In the unpreserved samples, the $Fe(II)/Fe_{Total}$ ratio is <0.03 for all determinations. This ratio in the unpreserved samples may be overestimated because of the possible formation of Fe(III)-phenanthroline complexes (Tamura et al., 1974). This interference is significant when the concentration of Fe(III) is >10 mg/L, a condition that was avoided during the partial extraction tests. In the preserved samples the average $Fe(II)/Fe_{Total}$ ratio is 0.83 (Figure 1). In most samples, concentrations of total iron and ferrous iron are in close agreement. Several other samples, in particular ED-14 and ED-15, were collected from a transitional redox zone so that a mixture of Fe(II) and Fe(III) found in these samples is reasonable. These data demonstrate that: i) the freezing procedure for preserving sample redox integrity is appropriate for iron-bearing phases; and, ii) samples containing ferrous iron, if left unpreserved, will undergo oxidation reactions that result in the conversion of Fe(II) to Fe(III) in the solid phase.

Similar results are observed for sulfur. In Figure 2 data are presented that show the concentration of acid-volatile sulfide in preserved and unpreserved sediment samples compared to total sulfur concentrations. Methods used for determining total sulfur and reduced sulfur partitioning are reported in Wilkin and Bischoff (2006). In the preserved set of samples, concentrations of acid-volatile sulfide range from 0.05 to 5.1 wt% or from about 10 to 79% of the total amount of sulfur contained in the samples. In contrast, the unpreserved samples have acid-volatile sulfide concentrations ranging from 0.01 to 0.18 wt%. Losses of acid-volatile sulfide concentrations range from 95 to 100% in the unpreserved samples. More detailed sulfur partitioning studies indicate that the balance of sulfur in the preserved samples is composed of mixedreducedandoxidizedspeciesincludingchromium-reducible sulfur, sulfate-sulfur, and minor quantities of organic-sulfur (Wilkin and Bischoff, 2006). Similarly to Fe(II), S(-II) is lost from samples that are left unpreserved.

A comparison was made between acid-volatile sulfide concentrations obtained in sediment samples that were thawed and dried in an anaerobic chamber and concentrations in freeze dried samples. Very good agreement was found between the two drying procedures (R = 0.953; n = 8). Freeze-drying may be advantageous for sample drying because low temperatures during lyophilization help avoid changes in labile components including the loss of volatile constituents (e.g., mercury, Muhaya et al., 1998), avoid aggregation of particles, and minimize oxidation reactions. A previous study showed, however, that freeze-drying was not effective for samples with low acid-volatile sulfide concentrations (Brumbaugh and Arms, 1996). At acid-volatile sulfide concentrations below

0.2 wt%, Brumbaugh and Arms (1996) noted reductions in concentrations following freeze-drying of up to 95%. They proposed that increases in sample surface area of freeze dried materials render such materials highly susceptible to air-oxidation. Hjorth (2004) also suggests that freeze-drying does not preserve the speciation pattern of major elements, trace metals, and sulfur in anoxic sediments as determined by a 3-step sequential extrac tion procedure. Although more data are needed, results available in the literature suggest that freeze-drying may not be an ideal approach for samples to be used in redox-sensitive solid-phase measurements;room-temperature drying in an anaerobic environment is preferred.

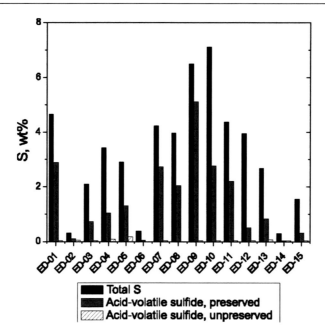

Figure 2. Comparison of solid-phase concentrations of total sulfur (wt%) and acid-volatile sulfide (wt%) in a series of sediments with and without preservation.

ARSENIC OXIDATION STATE

Arsenic may be present in the solid phase in either the As(V) or As(III) oxidation states, or as a mixture of As(V) and As(III). In general, as predicted from thermodynamic reasoning, As(V) is expected to dominate in oxygenated environments and As(III) is expected to dominate in suboxic to anoxic environments. Various mechanisms of arsenic mobilization and immobilization in the environment have been proposed including abiotic and microbially mediated redox processes. Determination of the oxidation state of arsenic in the solid phase is an important component of risk assessments and remediation strategies because both the toxicity and the geochemical mobility of arsenic are strongly dependent on its solid and aqueous phase speciation (e.g., Cullen and Reimer, 1989; Smedley and Kinniburgh, 2002).

Changes in the natural distribution of arsenic species in a sample collected from the field can come about due to several factors including chemical reactions with sample components, interactions with the container material, and microbial activity. All of these factors may in turn be affected by parameters such as temperature, light levels, and pH (Rowland et al., 2005). In this study, XANES spectra were collected to evaluate the oxidation state of arsenic in the preserved and unpreserved sediments. A discussion of data collection and data analysis methods relating to X-ray absorption spectroscopy is presented in a separate report

(U.S. Environmental Protection Agency, 2005). As an example, X-ray absorption spectra for preserved and unpreserved samples of ED-03 are shown in Figure 3. The preserved sample shows a single absorption maximum at about 11871 eV. This energy is characteristic of arsenic in the trivalent state dominant in the preserved sample. The unpreserved sample shows two features, a shoulder at 11871 eV and an absorption maximum at about 11874 eV. The second energy feature is characteristic of arsenic in the pentavalent state. Linear combination fitting of the measured spectra indicate that the unpreserved sample contains a mixture consisting of about 54% As(III) and 46% As(V).

Similar to iron and sulfur, the oxidation state of arsenic in the solid phase is highly dependent on the mode of sample preservation. Unless preserved, solid matrices containing arsenic in the trivalent state will likely oxidize to form arsenate. As an example, Bostick et al. (2004) documented arsenic oxidation artifacts encountered during spectroscopic measurements. In this study, sample freezing followed by sample preparation and analysis under an anoxic atmosphere was found to preserve the reduced arsenic oxidation state in solid samples. These findings are consistent with a recent study by Rowland et al. (2005). They noted substantial oxidation of solid-phase arsenic in unpreserved samples and that sandy matrices were particularly susceptible to arsenic oxidation. For sand-dominated samples, Rowland et al. (2005) recommend analysis within two or three weeks of sample collection to minimize oxidation artifacts. The issue of holding time was not specifically examined as an experimental variable in this study. Arsenic XANES data reported here were collected 5 months after sample collection, so over a 5-month period the arsenic oxidation state was maintained in the redox-preserved samples by freezing.

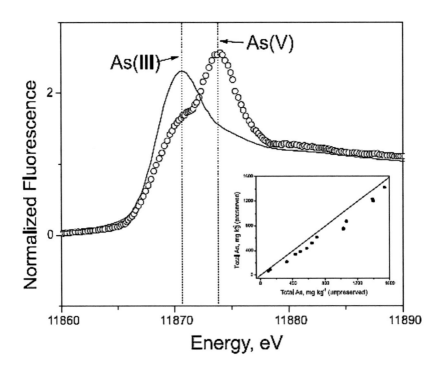

Figure 3. Arsenic K-edge XANES spectra for sample ED-03. The bold blue line is the spectrum collected from the preserved sample and the open circles show the spectrum collected from the unpreserved sample. Sample ED-03 contains a total arsenic concentration of 490 mg/kg; only As(III) is detected in the preserved sample. Inset shows the comparison between total arsenic concentrations in the preserved and unpreserved samples.

BATCH ADSORPTION TESTS

The capacity of soils, sediments, or aquifer solids to attenuate pollutants is often assessed by using batch-adsorption or static equilibrium tests (U.S.Environmental Protection Agency, 1992). It is reasonable to suspect that the results of such tests will depend on the ways in which sample matrices are preserved and handled after their collection. To examine the effects of preservation on batch-adsorption experiments, tests with zinc and arsenic were carried out using sub-samples of the preserved and unpreserved sediments. Zinc sorption onto the preserved samples was about 5 to 30 times greater than zinc sorption onto the unpreserved samples (Figure 4A). Interestingly, the reverse trend is evident for arsenic, i.e., the unpreserved samples are about 4 times more efficient in removing arsenic from solution compared to the preserved samples (Figure 4B). Figure 4 shows batch sorption data plotted in terms of the aqueous concentrations of zinc or arsenic in mg/L versus the solid-phase concentration of zinc or arsenic in mg/g, respectively. The solid-phase concentration of the inorganic contaminant is calculated based upon the dry sample mass used in the batch adsorption test and the time-dependent loss of zinc or arsenic concentrations from solution. The solid lines are the data fit to the Langmuir isotherm equation; the fitting equation is provided in the caption for Figure 4. The dashed lines in Figure 4 represent the linear distribution coefficient (K_d) for zinc and arsenic uptake by the unpreserved and preserved samples, respectively. Inthesetwo cases the linear K_d model would appear to be just as appropriate as the non-linear Langmuir model. Results of these tests, however, demonstrate that data collected in batch adsorption experiments are entirely dependent on how samples are preserved. Sample preservation would be just as important for column experiments. If solid-phase testing is used in the context of developing predictive models of contaminant transport and fate or for developing site remediation strategies, it is imperative that the solid-phase tests be carried out only on redox-preserved materials.

Arsenic is preferentially retained on the unpreserved sediment matrix. This behavior is likely due to the fact that both arsenite and arsenate are more favorably adsorbed by ferric oxyhydroxides or hydroxides present in the unpreserved samples as compared to ferrous sulfides that are present in the preserved samples. On the other hand, zinc is preferentially retained on the preserved sediment sample relative to the unpreserved sample. The high acid-volatile sulfide concentrations in the preserved samples provide reactive sulfide for precipitation of insoluble zinc sulfide (ZnS), which is a more effective process for removing zinc from solution than adsorption by ferric oxyhydroxides or hydroxides.

SUMMARY AND CONCLUSIONS

Unless preserved, samples collected from suboxic to anoxic environments should not be submitted for solid-phase tests to assess contaminant partitioning or for determining contaminant uptake capacity. Results of such tests on improperly preserved samples will be unrepresentative at best and misleading in the worst case.

The preservation method tested and recommended here for samples collected from suboxic to anoxic environments involves collection of samples followed by freezing (-18 °C), transporting frozen samples on dry ice, and laboratory processing of solids in an anaerobic glove box. This method was found to preserve the redox integrity of reduced iron- and sulfur-bearing compounds which are typically abundant redox-sensitive constituents in environmental samples.

The method is relatively simple and inexpensive to apply in the field compared to other possible methods of preservation that require liquid nitrogen or compressed gas cylinders containing nitrogen or argon. A selection of solid-phase measurements was carried out on preserved anoxic sediments collected from a contaminated wetland and compared to sample splits in which no preservation was adopted, i.e., the unpreserved samples were allowed to oxidize in air. The examples provided in this Issue Paper show that attention must be paid to sample preservation protocols, especially in site assessments that focus on the details of metal or metalloid partitioning to the solid matrix. Improper preservation practices prior to metal partitioning or batch adsorption tests may result in misleading data that are unrepresentative of site conditions. Changes in the oxidation state of iron and sulfur result in mineralogical changes that significantly impact contaminant behavior during characterization tests. Freezing was found to be an adequate method for preserving samples containing reduced iron, sulfur, and arsenic. When solid-phase tests such as metal speciation analyses, sequential extraction tests, or batch adsorption experiments are carried out on samples collected from anoxic environments, sample preparation and testing must be conducted in an oxygen-free atmosphere.

Although this study focused on a limited set of redox-sensitive elements from only one environment (i.e., freshwater sediment), it is reasonable to expect that the methods employed would be appropriate in other environmental media and for other redox-sensitive elements of interest (e.g., Mn, Se, U, V). Additional studies are needed to address redox preservation over a more complete range of contaminant types and environmental conditions. Other specific issues that require more study include an analysis of methods for preserving organic carbon fractions such as humic substances and an evaluation of storage times for specific redox-sensitive components.

EPA's Office of Research and Development is preparing a technical resource document for the application of monitored natural attenuation (MNA) to inorganic contaminants in ground water (see, e.g., Reisinger et al., 2005). The technical resource document presents a four-tiered analysis for assessing MNA as a viable remediation option for selected metal, metalloid, and radionuclide contaminants encountered in ground water. Components of the tiered approach include demonstrating contaminant sequestration mechanisms, estimating attenuation rates and the attenuation capacity of aquifer solids, and determining potential reversibility issues. All of these issues require samples that are representative of actual environmental conditions in order to evaluate MNA as a possible remedy for restoring ground water resources. Redox preservation of solids collected from the field will necessarily be a key component of MNA assessments for inorganic contaminants.

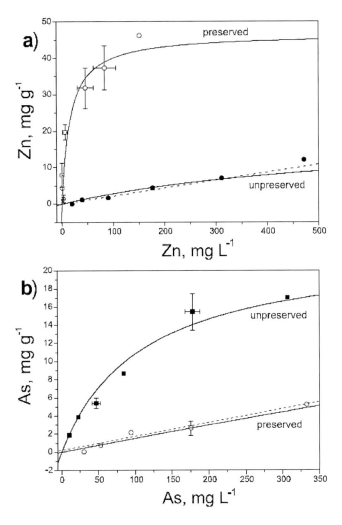

Figure 4. Adsorption isotherms of a) zinc and b) arsenic for preserved (blue) and unpreserved (black) sediment. Batch adsorption experiments were carried out using sample ED-10; pH of adsorption varied between 5.7 and 7.1. Solid lines show the fit to the Langmuir isotherm equation:

$$Q = Q_{max}\left(\frac{K_{ads}C}{1 + K_{ads}C}\right)$$

where Q is the concentration dependent sorption(mg/g), Q_{max} is the maximum possible sorption by the solid, C is the aqueous concentration of the sorbate(mg/L), and K_{ads} is the sorption constant (L/mg). Dashed lines show the fit to the linear adsorption model: $K_d = Q/C$.

NOTICE

The U.S. Environmental Protection Agency through its Office of Research and Development funded and managed the research described here. It has been subjected to the Agency's peer and administrative review and has been approved for publication as an EPA document. Mention of trade names or commercial products does not constitute endorsement or recommendation for use.

QUALITY ASSURANCE STATEMENT

All research projects making conclusions or recommendations based on environmental data and funded by the U.S. Environmental Protection Agency are required to participate in the Quality Assurance Program. This project was conducted under an approved Quality Assurance Project Plan. The procedures specified in this plan were used without exception. Information on the plan and documentation of the quality assurance activities and results are available from the Principal Investigator.

ACKNOWLEDGEMENTS

R. Ford and K. Scheckel are thanked for field and laboratory assistance. We also gratefully acknowledge the support provided by Shaw Environmental (Contract #68-C-03-097). This Issue Paper was reviewed by C. Stein (Gannett Fleming, Inc.), C. Cooper (Idaho National Laboratory), R. Ford (USEPA/ORD), and D. Frank (USEPA/Region 10);their comments and suggestions for improvement of the manuscript are greatly appreciated. Arsenic K-edge spectra were collected at beamline 20-BM at the Advanced Photon Source, Argonne National Laboratory (Argonne, IL). Use of the Advanced Photon Source is supported by the U. S. Department of Energy, Office of Science, Office of Basic Energy Sciences, under Contract #W-31-109-Eng-38. Advanced Photon Source research facilities at beamline 20-BM are also supported by the US DOE Office of Science Grant No. DEFG03-97ER45628, the University of Washington, a major facilities access grant from NSERC, Simon Fraser University and the Advanced Photon Source.

REFERENCES

Barcelona, M.J. and Holm, T.R.(1991). Oxidation-reductioncapacities of aquifer solids. *Environmental Science and Technology*, v. 25, p. 1565-1572.

Bostick, B.C., Chen, C., and Fendorf, S. (2004). Arsenite retention mechanisms within estuarine sediments of Pescadero, CA. *Environmental Science and Technology*, v. 38, p. 3299-3304.

Bottcher, G., Brumsack, H.-J., Heinrichs, H., and Pohlmann, M. (1997). A new high-pressure squeezing technique for pore fluid extraction from terrestrial soils. *Water, Air, and Soil Pollution*, v. 94, p. 289-296.

Brumbaugh, W.G. and Arms, J.W. (1996). Quality control considerations for the determination of acid-volatile sulfide and simultaneously extracted metals in sediments. *Environmental Toxicology and Chemistry*, v. 15, p. 282-285.

Bush, R.T. and Sullivan, L.A. (1997). Morphology and behaviour of greigite from a Holocene sediment in Eastern Australia. *Australian Journal of Soil Research*, v. 35, p. 853-861.

Carbonaro, R.F., Mahony, J.D., Walter, A.D., Halper, E.B., and DiToro, D.M. (2005). Experimental and modeling investigation of metal release from metal-spiked sediments. *Environmental Toxicology and Chemistry*, v. 24, p. 3007-3019.

Cullen, W.R. and Reimer, K.J. (1989). Arsenic speciation in the environment. *Chemical Reviews*, v. 89, p. 713-764.

Givelet, N., Le Roux, G., Cheburkin, A., Chen, B., Frank, J., Good-site, M.E., Kempter, H., Krachler, M., Noernberg, T., Rausch, N., Rheinberger, S., Roos-Barraclough, F., Sapkota, A., Scholz, C., and Shotyk, W. (2004). Suggested protocol for collecting, handling and preparing peat cores and peat samples for physical, chemical, mineralogical and isotopic analyses. *Journal of Environmental Monitoring*, v. 6, p. 481-492.

Harrington, J.M., Fendorf, S.E., Wielinga, B.W., and Rosenzweig, R.F. (1999). Response to comment on "Phase associations and mobilization of iron and trace elements in Coeur d'Alene Lake, Idaho". *Environmental Science and Technology*, v. 33, p. 203-204.

Hjorth, T. (2004). Effectsoffreeze-dryingonpartitioningpatternsof major elements and trace elements in lake sediments. *Analytica Chimica Acta*, v. 526, p. 95-102.

Mudroch, A. and Bourbonniere, R.A. (1994). Sediment preservation, processing, and storage. In Mudroch, A. and Macknight,

S.D. (eds.), *Handbook of Techniques for Aquatic Sediments Sampling*, Chapter 6, pp. 131-169. Lewis Publishers, Boca Raton, FL.

Muhaya, B.B.M., Leermakers, M., and Baeyens, W.(1998). Influence of sediment preservation on total mercury and methylmercury analyses. *Water, Air, and Soil Pollution*, v. 107, p. 277-288.

Phillips, D.H., Gu, B., Watson, D.B., and Roh, Y. (2003). Impact of sample preparation on mineralogical analysis of zero-valent iron reactive barrier materials. *Journal of Environmental Quality*, v. 32, p. 1299-1305.

Rapin, F., Tessier, A., Campbell, P.C.G., and Carignan, R. (1986). Potential artifacts in the determination of metal partitioning in sediments by a sequential extraction procedure. *Environmental Science and Technology*, v. 20, p. 836-841.

Reisinger, H.J., Burris, D.R., and Hering, J.G. (2005). Remediating subsurface arsenic contamination with monitored natural attenuation. *Environmental Science and Technology*, v. 39, p. 458A-464A.

Rowland, H.A.L., Gault, A.G., Charnock, J.M., and Polya, D.A. (2005). Preservation and XANES determination of the oxidation state of solid-phase arsenic in shallow sedimentary aquifers in Bengal and Cambodia. *Mineralogical Magazine*, v. 69, p. 825-839.

Smedley, P.L. and Kinniburgh, D.G.(2002). A review of the source, behaviour and distribution of arsenic in natural waters. *Applied Geochemistry*, v. 17, p. 517-568.

Tamura, H., Goto, K., Yotsuyanagi, T., and Nagayama, M. (1974). Spectrophotometric determination of iron(II) with 1,10-phenanthroline in the presence of large amounts of iron(III). *Talanta*, v. 21, p. 318-321.

U.S. Environmental Protection Agency (1992). Batch-type procedures for estimating soil adsorption of chemicals. USEPA Office of Solid Waste and Emergency Response, EPA/530/SW87/0066-F, Washington DC.

U.S. Environmental Protection Agency (2002). Workshop on monitoring oxidation-reduction processes for ground-water restoration. USEPA National Risk Management Research Laboratory, EPA/600/R-02/002, Cincinnati, OH.

U.S. Environmental Protection Agency (2003). Capstone report on the application, monitoring, and performance of permeable reactive barriers for ground-water remediation; Volume 1, Performance evaluations at two sites. USEPA National Risk Management Research Laboratory, EPA/600/R-03/045a, Cincinnati, OH.

U.S. Environmental Protection Agency (2005).   Field study of the fate of arsenic, lead and zinc at the ground water/surface water interface. USEPA National Risk Management Research Laboratory, EPA/600/R-05/161, Cincinnati, OH.

Wilkin, R.T. and Bischoff, K.J. (2006). Coulometric determination of total sulfur and reduced inorganic sulfur fractions in environmental samples. *Talanta*, v. 70, p. 766-773.

INDEX

D

E

F

T